Methods in Molecular Biology

Series Editor
John M. Walker
School of Life and Medical Sciences
University of Hertfordshire
Hatfield, Hertfordshire, AL10 9AB, UK

For further volumes:
http://www.springer.com/series/7651

Proteome Bioinformatics

Edited by

Shivakumar Keerthikumar and Suresh Mathivanan

*Department of Biochemistry and Genetics, La Trobe Institute for Molecular Science,
La Trobe University, Melbourne, VIC, Australia*

Humana Press

Editors
Shivakumar Keerthikumar
Department of Biochemistry and Genetics
La Trobe Institute for Molecular Science
La Trobe University
Melbourne, VIC, Australia

Suresh Mathivanan
Department of Biochemistry and Genetics
La Trobe Institute for Molecular Science
La Trobe University
Melbourne, VIC, Australia

ISSN 1064-3745 ISSN 1940-6029 (electronic)
Methods in Molecular Biology
ISBN 978-1-4939-6738-4 ISBN 978-1-4939-6740-7 (eBook)
DOI 10.1007/978-1-4939-6740-7

Library of Congress Control Number: 2016959985

Printed on acid-free paper

This Humana Press imprint is published by Springer Nature
The registered company is Springer Science+Business Media LLC
The registered company address is: 233 Spring Street, New York, NY 10013, U.S.A.

Preface

Recently, mass spectrometry (MS) instrumentation and computational tools have witnessed significant advancements. Thus, MS-based proteomics continuously improved the way proteins are identified and functionally characterized. This book covers the most recent proteomics techniques, databases, bioinformatics tools, and computational approaches that are used for the identification and functional annotation of proteins and their structure. The most recent proteomic resources widely used in the biomedical scientific community for storage and dissemination of data are discussed. In addition, specific MS/MS spectrum similarity scoring functions and their application in the field of proteomics, statistical evaluation of labeled comparative proteomics using permutation testing, and methods of phylogenetic analysis using MS data are also described in detail.

This edition includes recent cutting-edge technologies and methods for protein identification and quantification using tandem MS techniques. The reader gets the details of both experimental and computational methods and strategies in the identifications and functional annotation of proteins. Readers are expected to have basic bioinformatics and computational skills for a clear understanding of this book.

We hope the scope of this book is useful for researchers who are beginners as well as advanced in the field of proteomics. We are extremely grateful to our colleagues who contributed high-quality chapters to this book. We thank the Springer publishers for their support and are grateful to Professor Emeritus John Walker.

Melbourne, VIC, Australia

Shivakumar Keerthikumar
Suresh Mathivanan

Contents

Contributors

SEONG BEOM AHN • *Department of Biomedical Sciences, Faculty of Medicine and Health Sciences, Macquarie University, Sydney, NSW, Australia*
SUSHMA ANAND • *Department of Biochemistry and Genetics, La Trobe Institute for Molecular Science, La Trobe University, Melbourne, VIC, Australia*
CHING-SENG ANG • *The Bio21 Molecular Science and Biotechnology Institute, University of Melbourne, Parkville, VIC, Australia*
MARK S. BAKER • *Department of Biomedical Sciences, Faculty of Medicine and Health Sciences, Macquarie University, Sydney, NSW, Australia*
JOEL CHICK • *Department of Cell Biology, Harvard Medical School, Boston, MA, USA*
NAVEEN CHILAMKURTI • *Department of Computer Science and Information Technology, School of Engineering and Mathematical Sciences, La Trobe University, Bundoora, VIC, Australia*
DAVID CHISANGA • *Department of Computer Science and Information Technology, School of Engineering and Mathematical Sciences, La Trobe University, Bundoora, VIC, Australia*
BRETT COOKE • *Department of Chemistry and Biomolecular Sciences, Australian Proteome Analysis Facility, Macquarie University, Sydney, NSW, Australia*
JOSEPH CURSONS • *Systems Biology Laboratory, Melbourne School of Engineering, The University of Melbourne, Parkville, VIC, Australia; ARC Centre of Excellence in Convergent Bio-Nano Science and Technology, Melbourne School of Engineering, The University of Melbourne, Parkville, VIC, Australia*
DEBASIS DASH • *G.N. Ramachandran Knowledge Centre for Genome Informatics, CSIR-Institute of Genomics and Integrative Biology, Delhi, India*
MELISSA J. DAVIS • *Systems Biology Laboratory, Melbourne School of Engineering, The University of Melbourne, Parkville, VIC, Australia; Bioinformatics Division, Walter and Eliza Hall Institute of Medical Research, Parkville, VIC, Australia; Faculty of Medicine, Dentistry and Health Science, Department of Biochemistry and Molecular Biology, The University of Melbourne, Parkville, VIC, Australia*
KEVIN M. DOWNARD • *Prince of Wales Clinical School, University of New South Wales, Sydney, NSW, Australia; Lowy Cancer Research Centre, University of New South Wales, Sydney, NSW, Australia*
CRISELDA SANTAN FERNANDES • *Department of Chemistry and Biomolecular Sciences, Faculty of Science and Engineering, Macquarie University, Sydney, NSW, Australia*
HARSHA GOWDA • *Institute of Bioinformatics, Bangalore, India; YU-IOB Center for Systems Biology and Molecular Medicine, Yenepoya University, Mangalore, India*
PAUL HAYNES • *Faculty of Medicine and Health Sciences, Department of Chemistry and Biomolecular Sciences, Macquarie University, Sydney, NSW, Australia*
BEGOÑA HERAS • *Department of Biochemistry and Genetics, La Trobe Institute for Molecular Science, La Trobe University, Melbourne, VIC, Australia*
MICHELLE M. HILL • *The University of Queensland, Diamantina Institute, Translational Research Institute, Woolloongabba, QLD, Australia*

MOHAMMAD TAWHIDUL ISLAM • *Department of Chemistry and Biomolecular Sciences, Faculty of Science and Engineering, Macquarie University, Sydney, NSW, Australia*
SHIVAKUMAR KEERTHIKUMAR • *Department of Biochemistry and Genetics, La Trobe Institute for Molecular Science, La Trobe University, Melbourne, VIC, Australia*
DHIRENDRA KUMAR • *G.N. Ramachandran Knowledge Centre for Genome Informatics, CSIR-Institute of Genomics and Integrative Biology, Delhi, India*
SHIYONG MA • *Prince of Wales Clinical School, University of New South Wales, Sydney, NSW, Australia; Lowy Cancer Research Centre, University of New South Wales, Sydney, NSW, Australia*
JAYAKANTHAN MANNU • *Department of Plant Molecular Biology and Bioinformatics, Centre for Plant Molecular Biology and Biotechnology, Tamil Nadu Agricultural University, Coimbatore, India*
LENNART MARTENS • *Medical Biotechnology Center, VIB, Ghent, Belgium; Department of Biochemistry, Ghent University, Ghent, Belgium; Bioinformatics Institute Ghent, Ghent University, Ghent, Belgium*
SURESH MATHIVANAN • *Department of Biochemistry and Genetics, La Trobe Institute for Molecular Science, La Trobe University, Melbourne, VIC, Australia*
PREMENDU P. MATHUR • *School of Biotechnology, KIIT University, Bhubaneswar, India*
GEOFFREY J. MCLACHLAN • *School of Mathematics and Physics, The University of Queensland, St. Lucia, QLD, Australia*
MEHDI MIRZAEI • *Faculty of Medicine and Health Sciences, Department of Chemistry and Biomolecular Sciences, Faculty of Medicine and Health Sciences, Macquarie University, Sydney, NSW, Australia*
ABIDALI MOHAMEDALI • *Department of Chemistry and Biomolecular Sciences, Faculty of Science and Engineering, Macquarie University, Sydney, NSW, Australia; Department of Biomedical Sciences, Faculty of Medicine and Health Sciences, Macquarie University, Sydney, NSW, Australia*
MARK P. MOLLOY • *Faculty of Medicine and Health Sciences, Department of Chemistry and Biomolecular Sciences, Macquarie University, Sydney, NSW, Australia; Department of Chemistry and Biomolecular Sciences, Australian Proteome Analysis Facility, Macquarie University, Sydney, NSW, Australia*
ISHMAM NAWAR • *Department of Chemistry and Biomolecular Sciences, Faculty of Science and Engineering, Macquarie University, Sydney, NSW, Australia*
HIEN D. NGUYEN • *School of Mathematics and Physics, The University of Queensland, St. Lucia, QLD, Australia; The University of Queensland, Diamantina Institute, Translational Research Institute, Woolloongabba, QLD, Australia*
DANA PASCOVICI • *Department of Chemistry and Biomolecular Sciences, Australian Proteome Analysis Facility, Macquarie University, Sydney, NSW, Australia*
KRISHNA PATEL • *Institute of Bioinformatics, Bangalore, India; Amrita School of Biotechnology, Kollam, India*
MOHASHIN PATHAN • *Department of Biochemistry and Genetics, La Trobe Institute for Molecular Science, La Trobe University, Melbourne, VIC, Australia*
JASON J. PAXMAN • *Department of Biochemistry and Genetics, La Trobe Institute for Molecular Science, La Trobe University, Melbourne, VIC, Australia*
SWATHIK CLARANCIA PETER • *Department of Plant Molecular Biology and Bioinformatics, Centre for Plant Molecular Biology and Biotechnology, Tamil Nadu Agricultural University, Coimbatore, India*

Shoba Ranganathan • *Department of Chemistry and Biomolecular Sciences, Faculty of Science and Engineering, Macquarie University, Sydney, NSW, Australia*
Monisha Samuel • *Department of Physiology, Anatomy and Microbiology, School of Life Sciences, La Trobe University, Melbourne, VIC, Australia*
Manika Singh • *Institute of Bioinformatics, Bangalore, India; Amrita School of Biotechnology, Amrita Kollam, India*
Elien Vandermarliere • *Medical Biotechnology Center, VIB, Ghent, Belgium; Department of Biochemistry and Bioinformatics Institute Ghent, Ghent University, Ghent, Belgium*
Jason W.H. Wong • *Prince of Wales Clinical School, University of New South Wales, Sydney, NSW, Australia; Lowy Cancer Research Centre, University of New South Wales, Sydney, NSW, Australia*
Jemma X. Wu • *Department of Chemistry and Biomolecular Sciences, Australian Proteome Analysis Facility, Macquarie University, Sydney, NSW, Australia*
Yunqi Wu • *Faculty of Medicine and Health Sciences, Department of Chemistry and Biomolecular Sciences, Macquarie University, Sydney, NSW, Australia*
Amit Kumar Yadav • *G.N. Ramachandran Knowledge Centre for Genome Informatics, CSIR-Institute of Genomics and Integrative Biology, Delhi, India*
Şule Yilmaz • *Medical Biotechnology Center, VIB, Ghent, Belgium; Department of Biochemistry, Ghent University, Ghent, Belgium; Bioinformatics Institute Ghent, Ghent University, Ghent, Belgium*

Chapter 1

An Introduction to Proteome Bioinformatics

Shivakumar Keerthikumar

Abstract

High-throughput techniques are indispensable for aiding basic and translational research. Among them, recent advances in proteomics techniques have allowed biomedical researchers to characterize the proteome of multiple organisms. This remarkable advancement have been well complemented by proteome bioinformatics methods and tools. Proteome bioinformatics refers to the study and application of informatics in the field of proteomics. This chapter provides an overview of computational strategies, methods, and techniques reported in this book for bioinformatics analysis of protein data. An outline of many bioinformatics tools, databases, and proteomic techniques described in each of the chapters is given here.

Key words Proteomics, Proteins, Bioinformatics, Databases and computational tools

1 Introduction

In general, "bioinformatics" refers to the application of informatics/ computer science in the field of biology. The study of entire protein content of cell is referred to as the "proteome." The completion of the human genome project and the recent release of first draft of human proteome have generated massive amounts of genomic and proteomic data, respectively. Recent advancement in instrumentation have revolutionized the field of proteomics and the way in which thousands of proteins are identified, quantified, and characterized in a high-throughput fashion. To aid the scientific research community, various bioinformatics tools, databases, and computational algorithms were developed for storage, dissemination, and subsequent analysis of these proteomics data. This chapter outlines various techniques, resources, bioinformatics tools, and computational strategies widely employed in the field of proteomics. Based on the chapters contributed, the content of this book can be broadly categorized into different sections.

Shivakumar Keerthikumar and Suresh Mathivanan (eds.), *Proteome Bioinformatics*, Methods in Molecular Biology, vol. 1549, DOI 10.1007/978-1-4939-6740-7_1, © Springer Science+Business Media LLC 2017

2 Proteomic Databases and Repositories

Recent advancement in the high-resolution mass spectrometry based techniques have further increased the magnitude of proteomic data being generated. Proteomics community efforts have increased the dissemination and storage of these proteomics data in central repositories to aid scientific community for further downstream analysis. Chapter 2 describe general introduction about different online proteomics community resources to store raw and processed proteomic data and its application in the field of proteomics. Thousands of spectra generated using tandem mass spectrometry are assigned to proteins by using conventional sequence database search strategy. Chapter 3 covers different types of sequence databases and its role in specificity and sensitivity of protein identifications.

3 Proteomic Techniques and Computational Strategies Used in the Proteome Bioinformatics

There are various quantitation strategies employed using label-based and label-free methods for quantification of proteins. Chapter 4 describes the most commonly used quantitative proteomics techniques including stable isotope labeling methods using enzymatic, chemical, and metabolic strategies as well as label-free quantitation strategies. Using tandem mass tags (TMT), a type of labeled quantitative method, Chapter 5 details its sample processing, labeling, fractionation and data processing protocols in a stepwise fashion. Chapter 6 by Pathan et al. deals with fundamentals of protein identifications, different search methods, and rationale behind unassigned spectra. The main computational challenge remains in assigning thousands of spectra to their respective peptides and proteins. In general, different scoring functions have been developed and are used in assigning these experimental MS/MS spectrum to the theoretical MS/MS spectrum. Chapter 7 by Sule Yilmaz, Elien Vandermarliere, and Lennart Martens describes MS/MS spectrum similarity scoring functions and their applications in proteomics and assess their relative performance on sample data. Chapter 8 describes the details of targeted proteomics techniques using proteotypic peptides and its implications in the field of proteomics research. Chapter 9 describes statistical evaluation of labeled comparative proteomics profiling experiments using permutation test. This chapter covers various steps involved in permutation analysis with false discovery rate control using various computational strategies.

Besides conventional sequence database search method, de novo sequencing method is also used in spectral assignment which mainly benefits from identification of novel peptides which are missed in the traditional database search strategies. Chapter 10 describes a methodology to integrate de novo peptide sequencing using three commonly available software solutions in tandem, complemented by homology searching and manual validation of spectra for greater usage of de novo sequencing approach and for potentially increasing proteome coverage. Using de novo sequencing method along with proteolytic peptide mass maps and mapping of mass spectral data onto classical phylogenetic trees, Chapter 11 describes methods of phylogenetic analysis using protein mass spectrometry.

4 Functional Characterization of Proteins

Identifying thousands of proteins using tandem mass spectrometry also poses huge challenges in biological, functional, and structural interpretation of proteomics data. To gain functional insights of high throughput proteomic data, functional enrichment analysis based on gene ontology terms, biological pathways, and protein–protein interaction network is performed using various stand-alone tools and Web-based user friendly programs. Chapter 12 gives stepwise instructions of using these tools and Web-based resources mainly used in functional enrichment analysis. On the other hand, Chapter 13 describes functional annotation pipeline for those proteins with very little or no annotations available and known to be suitable for reconfirming data obtained from proteomics experiments.

An overview of basic network theory concepts and most commonly used protein–protein interaction network databases as well as computational tools used in the analysis of interaction network topology, biological modules and their visualization is described in Chapter 14. Statistical tests are usually performed to identify the significance of enriched or depleted proteins in these functional and interaction network analysis. However, Chapter 15 describes an alternative strategy and methodology to determine the statistical significance of network features using permutation testing.

Ultimate design of these computational tools, approaches, and resources, in this context, is to functionally and structurally characterize proteins. Determining three-dimensional structure of the proteins and identifying ligands to which they bind is an important step towards elucidating protein functions and advancement in X-ray crystallographic techniques has contributed to increasing number of protein structures. As a result various bioinformatics tools and resources have been developed to store and analyze these protein structures. Chapter 16 describes number of such freely available bioinformatics tools and databases used primarily for the analysis of protein structures determined using X-ray crystallographic techniques. One such application of these protein structure-determining tools and resources is described in Chapter 17.

Chapter 2

Proteomic Data Storage and Sharing

Shivakumar Keerthikumar and Suresh Mathivanan

Abstract

With the advent of high-throughput genomic and proteomic techniques, there is a massive amount of multidimensional data being generated and has increased several orders of magnitude. But the amount of data that is cataloged in the central repositories and shared publicly with the scientific community does not correlate the same rate at which the data is generated. Here, in this chapter, we discuss various proteomics data repositories that are freely accessible to the researchers for further downstream meta-analysis.

Key words Proteins, Peptides, Databases, False discovery rate, Cancer, Mass spectrometry

1 Introduction

The applications of mass spectrometry in identification and quantification of proteins in complex biological samples is rapidly evolving [1–3]. Recent technical advances in mass spectrometer to measure the abundance of proteins have further increased the amount of multidimensional data being generated [4]. As a result, significant interests have been created to characterize the proteome of many cell types and subcellular organelles [5–9]. There are three different layers of proteomic data that is generated using mass spectrometry-based techniques: raw data, peptide/protein data (also known as "result" or "peak list") and metadata. Raw data is basically a binary format file which most of the proteomic tools like MSConvert (http://proteowizard.sourceforge.net/tools.shtml) converts further into human readable formats such as mgf, XML, pkl, and txt files. Metadata contains experimental details, type of instruments, modifications and search engines/tools used [10]. In order to disseminate these different types of data to the scientific community, researchers have constantly thrived to develop central repository to store and share these humongous data [11–13].

Shivakumar Keerthikumar and Suresh Mathivanan (eds.), *Proteome Bioinformatics*, Methods in Molecular Biology, vol. 1549, DOI 10.1007/978-1-4939-6740-7_2, © Springer Science+Business Media LLC 2017

Here, we focus on publicly available free centralized resources that disseminate all kinds of proteomics data and tools which further aid in downstream analysis to gain new biological insights that benefit the scientific community.

2 Online Proteomics Community Resources

Currently, there are wide varieties of online resources (Table 1) that host different types of proteomics data at different level and software tools to further mine these data. The most commonly and widely used proteomic resources are discussed here.

2.1 PRoteomics IDEntifications (PRIDE) Database

The PRIDE database is most widely used centralized, publicly available proteomic repository which stores and manages all three different levels of proteomic data such as raw data, peak list file and metadata. The PRIDE database established at European Bioinformatics Institute, UK has a Web-based, user-friendly query and data submission system as well as documented application programming interface besides local installation [14]. Recently, the new PRIDE archival system (http://www.ebi.ac.uk/pride/archive/) has replaced the PRIDE database. The PRIDE archive system supports community recommended Proteomic Standard Initiative (PSI) data formats and is an active founding member of ProteomeXchange (PX) consortium (http://www.proteomicexchange.org/). The main concept behind such consortium is to standardize the mass spectrometry proteomics data and automate the sharing of these data between the repositories to benefit the end users [15].

The PRIDE archive system also stores many software tools such as PRIDE Inspector, PRIDE converter and PX submission tool to further streamline the data submission process and its visualization to aid scientific community. All these software tools including Web modules are developed in JAVA and are open source (https://code.google.com/archive/p/ebi-pride/). Besides funding agencies, many scientific journals such as *Nature Biotechnology*, *Proteomics*, *Molecular and Cellular Proteomics* and *Journal of Proteome Research* mandates submission of raw data and associated metadata to proteomics repository to support their publication which further elevated the public deposition of proteomics data. As a result, The PRIDE archive currently contains ~140 TBs size of data which constitutes 690 M spectra, 298 M and 66 M peptide and protein identification, respectively, spanning more than 500 different taxonomical identifiers.

2.2 PeptideAtlas

The PeptideAtlas (http://www.peptideatlas.org/) database is another freely available mass spectrometry derived proteomic data repository developed at Institute of Systems Biology, Seattle, USA.

Table 1
Overview of online proteomics resources

Database	Types of data stored	Link
PRIDE	Accepts Raw data, processed data and meta data	http://www.ebi.ac.uk/pride/archive/
Peptide Atlas	Accepts only Raw data and limited meta data	http://www.peptideatlas.org/
CPTAC	Allows download and dissemination of raw data, processed data and meta data relevant to cancer biospecimens collated through Proteomic Characterization centers (PCCs)	http://proteomics.cancer.gov/
Colorectal Cancer Atlas	Stores processed protein and peptide data after automatically analyzing the publicly available raw data from the proteomic repositories	http://www.colonatlas.org/
GPMDB	Stores processed protein and peptide data after automatically analyzing the publicly available raw data from the proteomic repositories. Supports data analysis	http://www.thegpm.org/
ProteomicsDB	Accepts Raw data, processed data and meta data. Allows download of raw data, processed protein and peptide data.	http://www.proteomicsdb.org/
Human Proteome Map	Allows download of processed protein and peptide data.	http://www.humanproteomemap.org/
Human Proteinpedia	Accepts processed and meta data.	http://www.humanproteinpedia.org/
Human Protein Atlas	Allows download of protein and RNA expression in normal and tumor tissues and cell types	http://www.proteinatlas.org/

Represents list of publicly available online proteomics resources and repositories discussed in this chapter

The PeptideAtlas accepts only spectra files either in the form of RAW, mzML or mzXML format and limited metadata. Once submitted, the raw spectra files are processed using standardized data processing pipeline known as Trans Proteomics Pipeline (TPP) [16] and stored in the SBEAMS (Systems Biology Experiment Analysis Management System)-Proteomics module. Further, peptides identified with high score are mapped to their respective genome sequence representing species/sample specific build [17, 18]. Currently, the PeptideAtlas has 19 organism specific build which includes many model organisms such as human, yeast, worms, mouse, fly, rat, horse, and zebrafish, for important sample groups such as plasma, brain, liver, lung, colon cancer, heart, kidney, and urine.

The PeptideAtlas, similar to the PRIDE archive system, is one of the founding members of PX consortium that implemented standardization of the mass spectrometry-based proteomics data and automate the sharing of proteomic data across different repositories. Another important feature of the PeptideAtlas is investigation of proteotypic peptides which are defined as peptides that can uniquely and unambiguously identify specific protein. Currently, users can search proteotypic peptides from three different organisms such as human, mouse, and yeast. Identification of such high scoring peptides would further serve as most possible targets for Selected Reaction Monitoring (SRM) approach [19]. The PeptideAtlas SRM Experiment Library (PASSEL) is a component of the PeptideAtlas project that is designed to enable submission, dissemination, and reuse of SRM experimental results from analysis of biological samples. The raw data submitted via PASSEL are automatically processed and stored into the database which can be further downloaded or accessed via web interface [20].

Further, the distinct peptides and its associated proteins identified from the user submitted raw data files using TPP tool can be further depicted graphically in Cytoscape [21] plugin implemented in the PeptideAtlas. Overall, the PeptideAtlas depicts the normalized outlook of the user submitted data which further aid in genome annotation of different organisms using mass spectrometry derived proteomic data.

2.3 CPTAC (Clinical Proteomic Tumor Analysis Consortium) Portal

The CPTAC data portal (http://proteomics.cancer.gov/) launched in August 2011 by National Cancer Institute (NCI) is a freely available, centralized public proteomic data repository collected by proteomic characterization centers for the CPTAC framework. The proteomic characterization center constitutes of five teams namely Broad Institute of MIT and Harvard/Fred Hutchinson Cancer Research Center, Johns Hopkins University, Pacific Northwest National Laboratory, Vanderbilt University, and Washington University/University of North Carolina. The proteome characterization center implements proteomics candidate developmental pipeline for further protein identification and its verification to serve as high value targets for clinically useful diagnostics. In addition, proteomic data from The Cancer Genome Atlas (TCGA) data portal (http://cancergenome.nih.gov/), xenograft models and other tissue datasets of well-characterized genome using standardized Common Data Analysis Pipeline are analyzed to increase the significance of the results. The CPTAC data portal hosts mass spectrometry data of cancer biospecimens such as breast, colorectal, and ovarian cancer as well as global profiling of post-translational modifications of tumor tissues and cancer cell lines which accounts to more than 6 TB data. The CPTAC data portal also hosts data from the Clinical Proteomic Technologies for Cancer Initiative from 2006 to 2011, which was mainly developed to address the pre-analytical and analytical variability issues that are major barriers in the field of proteomics. The major outcome of this program was the launch of the CPTAC data portal to understand the molecular basis of cancer using proteomic technology [22, 23].

2.4 Colorectal Cancer Atlas

Colorectal Cancer Atlas (http://www.colonatlas.org/) is web-based resource developed by integrating genomic and proteomic annotations identified precisely in colorectal cancer tissues and cell lines. It integrates heterogeneous data freely available in the public repositories, published articles [24] and in-house experimental data pertaining to quantitative and qualitative protein expression data obtained from variety of techniques such as mass spectrometry, western blotting, immunohistochemistry, confocal microscopy, immunoelectron microscopy, and fluorescence-activated cell sorting. Colorectal Cancer Atlas collates raw proteomic mass spectrometry and other proteomic experimental data specifically from colorectal cancer tissues and cell lines is processed using in-house pipeline. The proteins/peptides identified after <5 % FDR cutoff is stored in the backend database. Besides, mutation data largely obtained by large and small sequencing methods are also incorporated into the Colorectal Cancer Atlas database [13].

Currently, Colorectal Cancer Atlas hosts >62,000 protein identifications, >8.3 million MS/MS spectra, >13,000 colorectal cancer tissues and >209 cell lines. Further, Colorectal Cancer Atlas facilitate users to visualize these proteins identified in context of signaling pathways, protein–protein interactions, gene ontology terms, protein domains, and posttranslational modifications. Users can download the entire colorectal cancer data in tab-delimited format using the download page at http://colonatlas.org/download/.

2.5 Global Proteome Machine Database (GPMDB)

The Global Proteome Machine Database (http://www.thegpm.org/) is another open source mass spectrometry based proteomic repository, publicly available for the scientific community. The GPMDB periodically checks all the public proteomic repositories, downloads and reanalyzes the proteomic data using X! Tandem search engine. Besides, the users can also use spectral search engine called X! Hunter (http://xhunter.thegpm.org/) and proteotypic profiler called X! P3 (http://p3.thegpm.org/) [25] to analyze their data. The resultant peptide and protein lists after passing through the stringent automated quality test are stored into the backend database along with relevant metadata. Further, the results can be either viewed in the GPM website or downloaded through ftp or other interfaces. Besides, the users can also submit their spectra files in different formats such as mgf, mzXML, pkl, mzData, dta, and common (for only big and compressed files) to GPM via 'Search Data' option available in the website. The most frequently checked public repositories for the suitable new proteomic data for reanalysis includes Proteome Xchange/PRIDE, PeptideAtlas/PASSEL, MassIVE (http://www.massive.ucsd.edu/), Proteomics DB, The Chorus Project (http://chorusproject.org/), and iProX (http://www.iprox.org/).

Recently, at the time of writing this chapter, the GPMDB released an updated version of the GPM Personal Edition-Fury to replace the old venerable Cyclone version and upgraded to the latest version of X! Tandem (Version 2015.12.15, Vengeance) which features speedy assignment of PTMs. In addition, the human and mouse protein identification information in GPMDB has been summarized into a collection of spreadsheets known as GPMDB Guide to Human Proteome (GHP) and GPMDB Guide to Mouse Proteome (GMP), respectively. This guide contains information organized into separate spreadsheets for each chromosome as well as mitochondrial DNA and made available for download at ftp://ftp.thegpm.org/projects/annotation/human_protein_guide/ and ftp://ftp.thegpm.org/projects/annotation/proteome_protein_guide/.

2.6 ProteomicsDB

ProteomicsDB (http://www.proteomicsdb.org/) is a human centric proteomic data repository developed jointly by Technical University Munich (TUM) and company SAP SE (Walldorf, Germany) and SAP Innovation Center and Cellzome GmbH (GSK Company). ProteomicsDB, an in-memory database, configured with 2 TB of random access memory (RAM) and 160 central processor units (CPU), designed for real-time analysis of big proteomic data. ProteomicsDB assembles raw proteomic data files from public repositories such as PRIDE, PeptideAtlas, MassIVE, ProteomeXchange, and many other individual laboratories as well as from in-house experiments and reprocess the files using MaxQuant [26] and MASCOT [27] software packages. The proteins and peptides identified after passing through quality control steps including FDR filters are deposited into ProteomicsDB.

ProteomicsDB came into the limelight in 2014 with the release of draft human proteome map assembled using mass spectrometry experiments on human tissues, cell lines, body fluids as well as data from PTM studies and affinity purifications [3]. Currently, at the time of writing, ProteomicsDB contains protein evidence for 15,721 of the 19,629 protein coding genes which constitutes 80% coverage of human proteome. ProteomicsDB has a Web-based user-friendly interface through which users can search and download details of particular protein and peptide sequence via 'browse by proteins' and 'browse by chromosomes' options. Besides, users can submit their raw mass spectrometry data files, peak list files and metadata associated with it only after creating a user account in the ProteomicsDB. The secure URL link generated. At the time of writing, there were more than 569 registered users, 76 projects and more than 400 experiments accounting to 7 TB of data in ProteomicsDB.

2.7 Human Proteome Map (HPM)

The Human Proteome Map (HPM) (http://www.humanproteomemap.org/) was developed to represent the draft study of human proteome map. The HPM database hosts high-resolution mass spectrometry proteomic data representing 17 adult tissues, six primary hematopoietic cells, and seven fetal tissues resulting in >84% human proteome coverage. The mass spectrometry data was searched against Human RefSeq database (version 50 with common contaminants) using SEQUEST (http://fields.scripps.edu/sequest/) and MASCOT [27] search engines through Proteome Discoverer 1.3 platform (Thermo Scientific, Bremen, Germany). The peptides and proteins identified were represented as normalized spectral counts and for each peptide the high resolution MS/MS spectrum for the best scoring peptides can be visualized using Lorikeet JQuery plugin (http://uwpr.github.io/Lorikeet/). The results of the proteins and peptides can be queried and downloaded in the standard formats, but the databases currently do not support the submission of any new proteomic data [2].

2.8 Human Proteinpedia

Human Proteinpedia (http://humanproteinpedia.org/) [28, 29] was developed in 2008 [2] to facilitate the sharing and integration of human proteomic data. Besides, it allows scientific community to contribute and maintain protein annotations using protein distributed annotation system also known as PDAS. Further, protein annotations submitted by the users are mapped to individual proteins and made available using Human Protein Reference database (HPRD: http://www.hprd.org/) [30]. This allows the user to visualize experimentally validated protein–protein interaction networks, protein expressions in cell lines/tissues, post-translational modifications and subcellular localizations besides mass spectrometry derived peptides/proteins and spectral details.

Human Proteinpedia enables users to query at gene/protein level, by types of tissue expressions, posttranslational modifications, subcellular localizations, different mass spectrometer types, and experimental platforms. Using PDAS, the users are allowed to upload only processed data (peak list files) and meta-data containing experimental details into the back-end database either using normal or batch (for high-throughput data) upload system. The entire Human Proteinpedia data can be further downloaded freely by the scientific community at http://www.humanproteinpedia.org/download/ [31].

Currently, more than 240 different laboratories around the world has contributed proteomic data into Human Proteinpedia database which resulted in >4.8 M MS/MS spectra, >1.9 M peptide identifications, >150,000 protein expressions, >17,000 posttranslational modifications, >34,000 protein–protein interactions, and >2900 subcellular localizations from >2700 proteomic experiments.

2.9 Human Protein Atlas

The Human Protein Atlas (HPA: http://www.proteinatlas.org/) hosts expression and localization of majority of human protein coding genes based on both RNA and protein data. It was developed in 2005 as a large scale effort to quest where the proteins encoded by the human protein coding genes are expressed in the different tissues and cell types. Unlike other proteomic resources mainly depends on mass spectrometry based protein identifications, the HPA largely uses antibody based proteomics and transcriptomics profiling methods to locate and identify proteins in tissues and cell types. The transcriptomic data quantifies gene expression levels on different tissues and cell types while antibody based protein profiling methods characterize spatial cellular distribution for the corresponding proteins at different substructures and cell types of the tissues [32].

At the time of writing this chapter, the Human Protein Atlas version 14 known to contain RNA data for 99% and protein data for 86% of the predictive human genes and includes >11 million images with primary data from immunohistochemistry and immunofluorescence. The HPA contains >37,000 validate antibodies corresponding to 17,000 human protein coding genes collated from 46 human cell lines and tissue samples from 360 people (44 normal tissue types from 144 people and the 20 most common types of cancer from 216 people) [33].

Recently, tissue-based map of the human proteome data analyzed from 32 tissues and 47 cell lines using integrated OMICS approach is included in the Human Protein Atlas to further explore the expression pattern across the human body. In addition, global analysis of secreted and membrane proteins (secretome and membrane proteome), as well as an analysis of expression profiles for all proteins targeted by pharmaceutical drugs (druggable proteome) and protein implicated in cancer (cancer proteome) is integrated into the Human Protein Atlas [9].

3 Discussion

The amount of proteomics data being shared among the scientific community is still not well organized when compared to the humongous data that is being generated due to advancement in the proteomics field. The main reason for this can be attributed to the limited funding available for the maintenance of the database server, manpower, and other infrastructure. As a result, few of the efficient repositories such as NCBI Peptidome [34, 35] and Tranche [10] are completely discontinued largely due to funding constraints. In order to sustain and serve the growing scientific community database like the CHORUS (https://chorusproject.org/), a cloud based platform for storage, analysis and sharing of mass spectrometry data is charging users with certain amount of fees based on type of services required. We urge the continuous usage of these proteomic resources and willingness to share the proteomics data to the scientific community will only keep these resources alive and stable. Further, these proteomics resources would aid as important discovery tools in the field of biomedical research.

References

1. Mathivanan S (2014) Integrated bioinformatics analysis of the publicly available protein data shows evidence for 96% of the human proteome. J Proteomics Bioinform 07:041–049. doi:10.4172/jpb.1000301
2. Kim MS, Pinto SM, Getnet D, Nirujogi RS, Manda SS, Chaerkady R, Madugundu AK, Kelkar DS, Isserlin R, Jain S, Thomas JK, Muthusamy B, Leal-Rojas P, Kumar P, Sahasrabuddhe NA, Balakrishnan L, Advani J, George B, Renuse S, Selvan LD, Patil AH, Nanjappa V, Radhakrishnan A, Prasad S, Subbannayya T, Raju R, Kumar M, Sreenivasamurthy SK, Marimuthu A, Sathe GJ, Chavan S, Datta KK, Subbannayya Y, Sahu A, Yelamanchi SD, Jayaram S, Rajagopalan P, Sharma J, Murthy KR, Syed N, Goel R, Khan AA, Ahmad S, Dey G, Mudgal K, Chatterjee A, Huang TC, Zhong J, Wu X, Shaw PG, Freed D, Zahari MS, Mukherjee KK, Shankar S, Mahadevan A, Lam H, Mitchell CJ, Shankar SK, Satishchandra P, Schroeder JT, Sirdeshmukh R, Maitra A, Leach SD, Drake CG, Halushka MK, Prasad TS, Hruban RH, Kerr CL, Bader GD, Iacobuzio-Donahue CA, Gowda H, Pandey A (2014) A draft map of the human proteome. Nature 509(7502):575–581. doi:10.1038/nature13302
3. Wilhelm M, Schlegl J, Hahne H, Moghaddas Gholami A, Lieberenz M, Savitski MM, Ziegler E, Butzmann L, Gessulat S, Marx H, Mathieson T, Lemeer S, Schnatbaum K, Reimer U, Wenschuh H, Mollenhauer M, Slotta-Huspenina J, Boese JH, Bantscheff M, Gerstmair A, Faerber F, Kuster B (2014) Mass-spectrometry-based draft of the human proteome. Nature 509(7502):582–587. doi:10.1038/nature13319
4. Lesur A, Domon B (2015) Advances in high-resolution accurate mass spectrometry application to targeted proteomics. Proteomics 15(5-6):880–890. doi:10.1002/pmic.201400450
5. Keerthikumar S, Gangoda L, Liem M, Fonseka P, Atukorala I, Ozcitti C, Mechler A, Adda CG, Ang CS, Mathivanan S (2015) Proteogenomic analysis reveals exosomes are more oncogenic than ectosomes. Oncotarget 6:15375–15396
6. Onjiko RM, Moody SA, Nemes P (2015) Single-cell mass spectrometry reveals small molecules that affect cell fates in the 16-cell embryo. Proc Natl Acad Sci U S A 112(21):6545–6550. doi:10.1073/pnas.1423682112
7. Lydic TA, Townsend S, Adda CG, Collins C, Mathivanan S, Reid GE (2015) Rapid and comprehensive 'shotgun' lipidome profiling of colorectal cancer cell derived exosomes. Methods 87:83–95. doi:10.1016/j.ymeth.2015.04.014
8. Habuka M, Fagerberg L, Hallstrom BM, Ponten F, Yamamoto T, Uhlen M (2015) The urinary bladder transcriptome and proteome defined by transcriptomics and antibody-based profiling. PLoS One 10(12):e0145301. doi:10.1371/journal.pone.0145301
9. Uhlen M, Fagerberg L, Hallstrom BM, Lindskog C, Oksvold P, Mardinoglu A, Sivertsson A, Kampf C, Sjostedt E, Asplund A, Olsson I, Edlund K, Lundberg E, Navani S, Szigyarto CA, Odeberg J, Djureinovic D, Takanen JO, Hober S, Alm T, Edqvist PH, Berling H, Tegel H, Mulder J, Rockberg J, Nilsson P, Schwenk JM, Hamsten M, von Feilitzen K, Forsberg M, Persson L, Johansson F, Zwahlen M, von Heijne G, Nielsen J, Ponten F (2015) Proteomics tissue-based map of the human proteome. Science 347(6220):1260419. doi:10.1126/science.1260419

10. No Authors Listed (2012) A home for raw proteomics data. Nat Methods 9(5):419
11. Keerthikumar S, Chisanga D, Ariyaratne D, Al Saffar H, Anand S, Zhao K, Samuel M, Pathan M, Jois M, Chilamkurti N, Gangoda L, Mathivanan S (2016) ExoCarta: a Web-based compendium of exosomal cargo. J Mol Biol 428(4):688–692. doi:10.1016/j.jmb.2015.09.019
12. Keerthikumar S, Raju R, Kandasamy K, Hijikata A, Ramabadran S, Balakrishnan L, Ahmed M, Rani S, Selvan LD, Somanathan DS, Ray S, Bhattacharjee M, Gollapudi S, Ramachandra YL, Bhadra S, Bhattacharyya C, Imai K, Nonoyama S, Kanegane H, Miyawaki T, Pandey A, Ohara O, Mohan S (2009) RAPID: resource of Asian primary immunodeficiency diseases. Nucleic Acids Res 37(Database issue):D863–D867. doi:10.1093/nar/gkn682
13. Chisanga D, Keerthikumar S, Pathan M, Ariyaratne D, Kalra H, Boukouris S, Mathew NA, Saffar HA, Gangoda L, Ang CS, Sieber OM, Mariadason JM, Dasgupta R, Chilamkurti N, Mathivanan S (2016) Colorectal cancer atlas: an integrative resource for genomic and proteomic annotations from colorectal cancer cell lines and tissues. Nucleic Acids Res 44(D1):D969–D974. doi:10.1093/nar/gkv1097
14. Vizcaino JA, Cote RG, Csordas A, Dianes JA, Fabregat A, Foster JM, Griss J, Alpi E, Birim M, Contell J, O'Kelly G, Schoenegger A, Ovelleiro D, Perez-Riverol Y, Reisinger F, Rios D, Wang R, Hermjakob H (2013) The PRoteomics IDEntifications (PRIDE) database and associated tools: status in 2013. Nucleic Acids Res 41(Database issue):D1063–D1069. doi:10.1093/nar/gks1262
15. Vizcaino JA, Csordas A, Del-Toro N, Dianes JA, Griss J, Lavidas I, Mayer G, Perez-Riverol Y, Reisinger F, Ternent T, Xu QW, Wang R, Hermjakob H (2016) 2016 update of the PRIDE database and its related tools. Nucleic Acids Res 44(D1):D447–D456. doi:10.1093/nar/gkv1145
16. Deutsch EW, Mendoza L, Shteynberg D, Slagel J, Sun Z, Moritz RL (2015) Trans-proteomic pipeline, a standardized data processing pipeline for large-scale reproducible proteomics informatics. Proteomics Clin Appl 9(7-8):745–754. doi:10.1002/prca.201400164
17. Farrah T, Deutsch EW, Hoopmann MR, Hallows JL, Sun Z, Huang CY, Moritz RL (2013) The state of the human proteome in 2012 as viewed through PeptideAtlas. J Proteome Res 12(1):162–171. doi:10.1021/pr301012j
18. Vizcaino JA, Foster JM, Martens L (2010) Proteomics data repositories: providing a safe haven for your data and acting as a springboard for further research. J Proteomics 73(11):2136–2146. doi:10.1016/j.jprot.2010.06.008
19. Pan S, Aebersold R, Chen R, Rush J, Goodlett DR, McIntosh MW, Zhang J, Brentnall TA (2009) Mass spectrometry based targeted protein quantification: methods and applications. J Proteome Res 8(2):787–797. doi:10.1021/pr800538n
20. Farrah T, Deutsch EW, Kreisberg R, Sun Z, Campbell DS, Mendoza L, Kusebauch U, Brusniak MY, Huttenhain R, Schiess R, Selevsek N, Aebersold R, Moritz RL (2012) PASSEL: the PeptideAtlas SRMexperiment library. Proteomics 12(8):1170–1175. doi:10.1002/pmic.201100515
21. Shannon P, Markiel A, Ozier O, Baliga NS, Wang JT, Ramage D, Amin N, Schwikowski B, Ideker T (2003) Cytoscape: a software environment for integrated models of biomolecular interaction networks. Genome Res 13(11):2498–2504. doi:10.1101/gr.1239303
22. Ellis MJ, Gillette M, Carr SA, Paulovich AG, Smith RD, Rodland KK, Townsend RR, Kinsinger C, Mesri M, Rodriguez H, Liebler DC, Clinical Proteomic Tumor Analysis C (2013) Connecting genomic alterations to cancer biology with proteomics: the NCI Clinical Proteomic Tumor Analysis Consortium. Cancer Discov 3(10):1108–1112. doi:10.1158/2159-8290.CD-13-0219
23. Edwards NJ, Oberti M, Thangudu RR, Cai S, McGarvey PB, Jacob S, Madhavan S, Ketchum KA (2015) The CPTAC data portal: a resource for cancer proteomics research. J Proteome Res 14(6):2707–2713. doi:10.1021/pr501254j
24. Mathivanan S, Ji H, Tauro BJ, Chen YS, Simpson RJ (2012) Identifying mutated proteins secreted by colon cancer cell lines using mass spectrometry. J Proteomics 76:141–149. doi:10.1016/j.jprot.2012.06.031
25. Craig R, Cortens JP, Beavis RC (2005) The use of proteotypic peptide libraries for protein identification. Rapid Commun Mass Spectrom 19(13):1844–1850. doi:10.1002/rcm.1992
26. Cox J, Mann M (2008) MaxQuant enables high peptide identification rates, individualized p.p.b.-range mass accuracies and proteome-wide protein quantification. Nat Biotechnol 26(12):1367–1372. doi:10.1038/nbt.1511
27. Perkins DN, Pappin DJ, Creasy DM, Cottrell JS (1999) Probability-based protein

identification by searching sequence databases using mass spectrometry data. Electrophoresis 20(18):3551–3567. doi:10.1002/(SICI)1522-2683(19991201)20:18<3551::AID-ELPS3551>3.0.CO;2-2

28. Mathivanan S, Ahmed M, Ahn NG, Alexandre H, Amanchy R, Andrews PC, Bader JS, Balgley BM, Bantscheff M, Bennett KL, Bjorling E, Blagoev B, Bose R, Brahmachari SK, Burlingame AS, Bustelo XR, Cagney G, Cantin GT, Cardasis HL, Celis JE, Chaerkady R, Chu F, Cole PA, Costello CE, Cotter RJ, Crockett D, DeLany JP, De Marzo AM, DeSouza LV, Deutsch EW, Dransfield E, Drewes G, Droit A, Dunn MJ, Elenitoba-Johnson K, Ewing RM, Van Eyk J, Faca V, Falkner J, Fang X, Fenselau C, Figeys D, Gagne P, Gelfi C, Gevaert K, Gimble JM, Gnad F, Goel R, Gromov P, Hanash SM, Hancock WS, Harsha HC, Hart G, Hays F, He F, Hebbar P, Helsens K, Hermeking H, Hide W, Hjerno K, Hochstrasser DF, Hofmann O, Horn DM, Hruban RH, Ibarrola N, James P, Jensen ON, Jensen PH, Jung P, Kandasamy K, Kheterpal I, Kikuno RF, Korf U, Korner R, Kuster B, Kwon MS, Lee HJ, Lee YJ, Lefevre M, Lehvaslaiho M, Lescuyer P, Levander F, Lim MS, Lobke C, Loo JA, Mann M, Martens L, Martinez-Heredia J, McComb M, McRedmond J, Mehrle A, Menon R, Miller CA, Mischak H, Mohan SS, Mohmood R, Molina H, Moran MF, Morgan JD, Moritz R, Morzel M, Muddiman DC, Nalli A, Navarro JD, Neubert TA, Ohara O, Oliva R, Omenn GS, Oyama M, Paik YK, Pennington K, Pepperkok R, Periaswamy B, Petricoin EF, Poirier GG, Prasad TS, Purvine SO, Rahiman BA, Ramachandran P, Ramachandra YL, Rice RH, Rick J, Ronnholm RH, Salonen J, Sanchez JC, Sayd T, Seshi B, Shankari K, Sheng SJ, Shetty V, Shivakumar K, Simpson RJ, Sirdeshmukh R, Siu KW, Smith JC, Smith RD, States DJ, Sugano S, Sullivan M, Superti-Furga G, Takatalo M, Thongboonkerd V, Trinidad JC, Uhlen M, Vandekerckhove J, Vasilescu J, Veenstra TD, Vidal-Taboada JM, Vihinen M, Wait R, Wang X, Wiemann S, Wu B, Xu T, Yates JR, Zhong J, Zhou M, Zhu Y, Zurbig P, Pandey A (2008) Human proteinpedia enables sharing of human protein data. Nat Biotechnol 26(2):164–167. doi:10.1038/nbt0208-164

29. Kandasamy K, Keerthikumar S, Goel R, Mathivanan S, Patankar N, Shafreen B, Renuse S, Pawar H, Ramachandra YL, Acharya PK, Ranganathan P, Chaerkady R, Keshava Prasad TS, Pandey A (2009) Human proteinpedia: a unified discovery resource for proteomics research. Nucleic Acids Res 37 (Database issue):D773–D781. doi:10.1093/nar/gkn701

30. Keshava Prasad TS, Goel R, Kandasamy K, Keerthikumar S, Kumar S, Mathivanan S, Telikicherla D, Raju R, Shafreen B, Venugopal A, Balakrishnan L, Marimuthu A, Banerjee S, Somanathan DS, Sebastian A, Rani S, Ray S, Harrys Kishore CJ, Kanth S, Ahmed M, Kashyap MK, Mohmood R, Ramachandra YL, Krishna V, Rahiman BA, Mohan S, Ranganathan P, Ramabadran S, Chaerkady R, Pandey A (2009) Human protein reference database—2009 update. Nucleic Acids Res 37 (Database issue):D767–D772. doi:10.1093/nar/gkn892

31. Muthusamy B, Thomas JK, Prasad TS, Pandey A (2013) Access guide to human proteinpedia. Curr Protoc Bioinformatics 1:121. doi:10.1002/0471250953.bi0121s41

32. Uhlen M, Oksvold P, Fagerberg L, Lundberg E, Jonasson K, Forsberg M, Zwahlen M, Kampf C, Wester K, Hober S, Wernerus H, Bjorling L, Ponten F (2010) Towards a knowledge-based human protein Atlas. Nat Biotechnol 28(12):1248–1250. doi:10.1038/nbt1210-1248

33. Marx V (2014) Proteomics: an atlas of expression. Nature 509(7502):645–649. doi:10.1038/509645a

34. Slotta DJ, Barrett T, Edgar R (2009) NCBI peptidome: a new public repository for mass spectrometry peptide identifications. Nat Biotechnol 27(7):600–601. doi:10.1038/nbt0709-600

35. Csordas A, Wang R, Rios D, Reisinger F, Foster JM, Slotta DJ, Vizcaino JA, Hermjakob H (2013) From peptidome to PRIDE: public proteomics data migration at a large scale. Proteomics 13(10-11):1692–1695. doi:10.1002/pmic.201200514

Chapter 3

Choosing an Optimal Database for Protein Identification from Tandem Mass Spectrometry Data

Dhirendra Kumar, Amit Kumar Yadav, and Debasis Dash

Abstract

Database searching is the preferred method for protein identification from digital spectra of mass to charge ratios (m/z) detected for protein samples through mass spectrometers. The search database is one of the major influencing factors in discovering proteins present in the sample and thus in deriving biological conclusions. In most cases the choice of search database is arbitrary. Here we describe common search databases used in proteomic studies and their impact on final list of identified proteins. We also elaborate upon factors like composition and size of the search database that can influence the protein identification process. In conclusion, we suggest that choice of the database depends on the type of inferences to be derived from proteomics data. However, making additional efforts to build a compact and concise database for a targeted question should generally be rewarding in achieving confident protein identifications.

Key words Shotgun proteomics, Peptide identification, Database size, Proteogenomics, neXtProt

1 Introduction

Comprehensive characterisation of proteome, the cellular workforce of an organism is important to understand the underlying biological phenomena and processes. Modern advances in ionization of biomolecules, multidimensional sample separation and mass spectrometry (MS) instrumentation have made shotgun proteomics the most popular approach to profile proteomes from biological samples in a high-throughput manner. During sample preparation proteins are isolated, digested into peptides with trypsin or other proteases, fractionated to reduce the complexity, and then injected in a mass spectrometer [1]. Digested peptides are ionized before flying inside a mass spectrometer either by electrospray ionization (ESI) [2] or matrix-assisted laser desorption ionization (MALDI) [3]. Often the detection of m/z of charged peptide ions is followed by fragmentation either by collision induced dissociation (CID) [4]or high-energy collision dissociation (HCD) or electron transfer dissociation (ETD) [5] to generate fragments due to bond

Shivakumar Keerthikumar and Suresh Mathivanan (eds.), *Proteome Bioinformatics*, Methods in Molecular Biology, vol. 1549, DOI 10.1007/978-1-4939-6740-7_3,

breakage along the peptide backbone. The set of m/z values and intensities of parent peptide along with its associated fragment ions represents one tandem (MS/MS) mass spectrum which is further utilized to identify peptide.

The interpretation of peptide sequences for thousands of spectra from one MS/MS run is generally carried out using either de novo, tag assisted or database search method. De novo approaches to interpret peptide or its partial tag from an MS/MS spectrum rely on accurate estimation of mass differences between m/z peaks and their correspondence to amino acid masses [6]. Although promising for infinite search space to decipher encrypted sequence in m/z values, it suffers from low resolution, low sensitivity, and partial coverage in peptide detection [7, 8]. Thus these methods are not viable for high-throughput proteomics. Instead, database search approaches are more popular to infer peptide and proteins from MS/MS data owing to their ease of automation. In database search method, spectra are searched against a protein database which represents biological protein sequences that might be present in the sample. Each protein in the database is theoretically digested into peptides following the cleavage rules of protease used in the experiment. Similar to the experimental process, theoretical mass spectra for these peptide sequences are simulated based on the fragmentation pattern specific to the dissociation method or instrument. These theoretical peptide spectra are compared with each experimental spectrum. The peptide which best explains the experimental spectrum also known as peptide spectrum match (PSM), is retained for further analysis [9]. To estimate the fraction of possible false matches due to random chance, multiple hypothesis testing is applied to the entire list of PSMs. For this, decoy database search based false discovery rate (FDR) estimation is routinely followed method [10, 11]. In this method, spectra are searched against a target database representing biological protein sequences and a decoy database containing all decoy or false proteins. The PSMs score distribution of decoy database search allows estimation of false positive fraction in PSMs assigned from target database. The FDR corrected list of PSMs leads to the list of peptides and proteins expressed in the sample. A schematic workflow of shotgun proteomics experiment and data analysis is presented in Fig. 1. Protein identification is an important step when quantitative changes in different biological samples or states are measured. The quantification of different proteins is dependent on peptide detection and thus the factors affecting peptide discovery would also impact the MS based quantitation of proteins. Another dimension of proteomic studies is to identify posttranslational modifications (PTMs) which contribute both dynamicity and diversity to the proteome. In the database search method anticipated PTMs can be discovered by defining the mass shift and amino acid specificity caused by these modifications. Defining PTMs during the

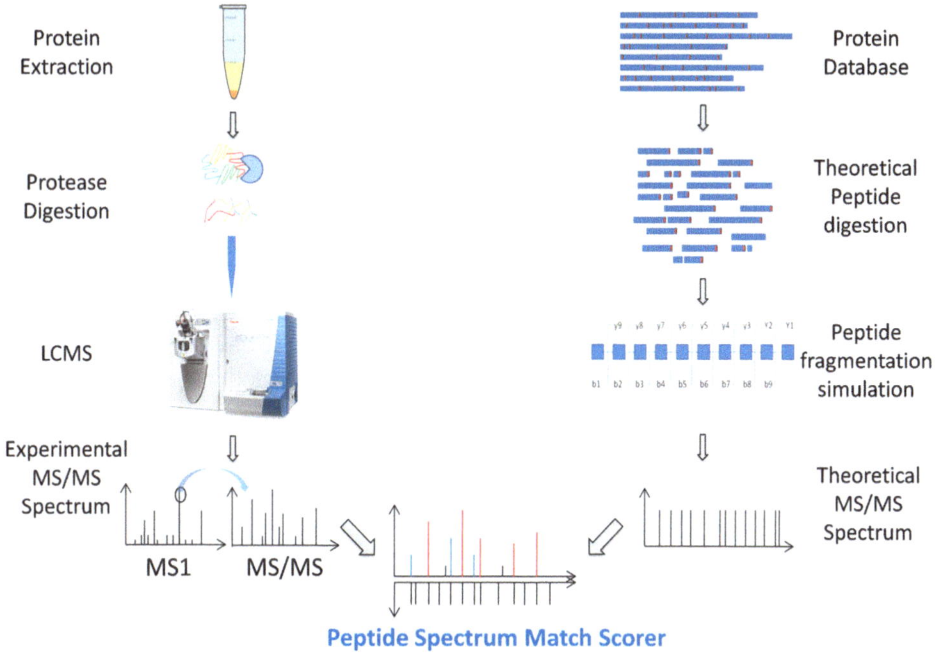

Fig. 1 Schematic workflow of protein detection from mass spectrometry based shotgun proteomics

search significantly alters the search space and thus influences the protein discovery.

Significance of the search database on proteomic studies can be understood by the fact that list of peptides and proteins vary significantly between different database searches. Further, different parameters applied change the effective search space, making the choice of database an important consideration. Thus it is important to understand which database would be optimal to maximize the protein discovery without increasing false positives.

2 Databases for Protein Discovery

Biological sequence information in the form of genome, transcriptome, and proteome can be retrieved from various global Web portals. Few resources like NCBI-RefSeq and UniProtKB host entire set of non redundant protein sequences, annotated or predicted and stored in the form of FASTA flat files. On the other hand, SwissProt, a small subset of UniProt comprises of only the confident set of proteins, the biological existence of which has been manually curated. UniProt provides the information on proteomes for 8975 organisms of which 2583 are reference proteomes (27/7/2015, http://www.uniprot.org/proteomes/). There are dedicated Web resources for various biologically important organisms as well. For example, neXtProt is a Web portal that annotates human proteins and isoforms for their levels of experimental observation [12]. GENCODE, a human gene annotation database also provides a list of protein coding human transcripts [13]. Genome resources like Ensembl and UCSC also provide proteomes for various eukaryotic and prokaryotic model organisms. Dedicated proteome resources for prokaryotes include Tuberculist for *Mycobacterium tuberculosis* (Mtb), EcoCyc and EcoGene for model bacterium *Escherichia coli*, and many others exist, from where organism-specific curated proteomes can be downloaded.

Given a proteomics dataset generated from an organism, the researcher has many options to select the search database from. However, in most cases the choice of database is arbitrary. For example, if one has proteomics data generated for a human cancer cell line, it can be searched against either of the databases like NCBI-RefSeq, UniProtKB, SwissProt, reference human proteomes from Ensebml/UniProt/NCBI or manually curated neXtProt. There are significant differences among these databases both in terms of size and content and thus identified protein list would vary amongst the searches. Similarly, if the data is generated for a bacterium, it can be searched against: (1) one of the reference proteomes from different sources, (2) all proteins known for the genus, (3) for the entire bacterial super-kingdom, or (4) entire SwissProt. Which of these searches will provide the optimal or maximal results is a difficult question to answer. Moreover, maximal may not always be optimal. Searching large datasets may result in higher number of hits, many of which may be random in nature. However, we discuss below the major factors which should be considered before deciding about the search database to achieve the optimal results.

2.1 Databases and Effect of Databases on Protein Discovery

2.1.1 Database Size Influences the Search Time and Results

One of the most easily distinguishable attribute of these search databases is the number of proteins they contain also referred to as the database size. The size of the database determines both the time complexity as well as the number of identified peptides from the searches. Figure 2 presents a size comparison of major proteomics search databases. While the global proteomes like NCBI-RefSeq or UniProtKB present comprehensive search space they are enormous when compared to reference proteomes. SwissProt on the other hand has a manageable search space. However, due to

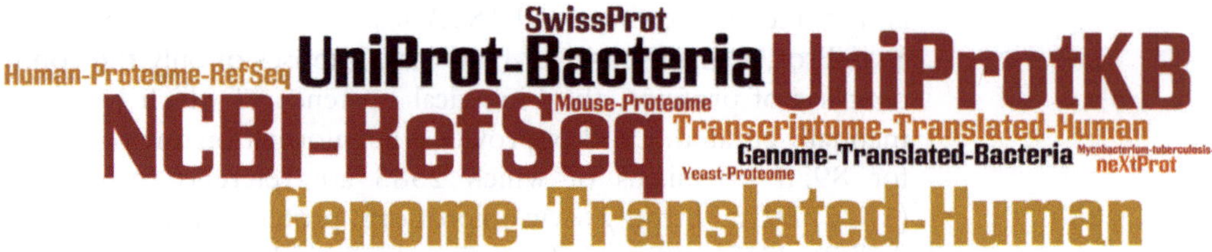

Fig. 2 Common proteomics search databases. *Font size* for each database name reflects database size

the manual curation of the database entries, comprehensiveness of the search space for non-model organisms and as-yet-unobserved proteins is debatable. The time complexity for proteomics database search increases linearly with increase in database size [8]. Thus it can be estimated that MS/MS data searching against global databases might consume ≈700× more time than the reference human proteome database.

2.1.2 Search Parameters Alter the Effective Search Space

Various search parameters reshape the search space from the original database. Important ones are listed below.

Precursor tolerance defines the expected limit of difference between experimental and theoretical peptide mass. In the search process, it determines the number of candidate peptides for a given spectrum. More the candidate peptides, more the number of comparisons between theoretical and experimental spectrum and thus more time it will require to determine the best scoring PSM for any given spectrum. Thus, increase in precursor tolerance would also increase the search time by expanding the effective search space.

Missed cleavages are sites where protease "missed" or did not cut during hydrolytic cleavage. The number of possible proteolytic peptides to consider for each protein increases with the higher number of missed cleavages and thus increasing candidate peptides per spectrum and search time. For example a theoretical cleavage of 479 aa long AKT3 protein produces 29, 86, 148, and 209 peptides (>7 aa) for 0, 1, 2, and 3 missed cleavages, respectively, indicating steep increase in possible candidates.

Posttranslational modifications (PTM) are commonly defined while searching the tandem mass spectrometry data as they signify the dynamic regulation of protein function and biological states. PTMs are generally searched as variable modifications due to their temporal nature. Variable modifications in search mean that they may or may not be present at a site. Therefore, various different modified peptide possibilities might exist and all these need to be considered while generating theoretical spectra for comparison with experimental spectrum. For example, if a peptide posses seven modifiable sites, total 99 (1 + 7 + 21 + 35 + 35) modified versions of this peptide are possible containing 0–4 sites as modified. Hence, it is estimated that search space increases exponentially with increase in PTMs defined in searches [8].

While searching MS/MS data, actual search space for each spectrum is determined by these search parameters and magnify database size many folds. Thus while determining an appropriate database not only the database size but the search parameters should also be considered.

2.1.3 Variability Between Databases

Another dimension which needs to be taken into consideration before deciding on the search database is its content in terms of protein sequences. Proteome set downloaded from different sources also vary in their protein content. For instance, human reference proteome can be retrieved from NCBI, UniProt, Ensembl, or Human Proteome Organization (HUPO) promoted neXtProt [14]. However, there are significant differences among these. While NCBI-RefSeq (Release 69) human proteome contains 72,123 protein sequences, reference proteomes from UniProt (27/7/2015), Ensembl (Release 78), and neXtProt (19/9/2014) contain 69,693, 99,436, and 41,038 proteins, respectively. Sequence similarity based comparison between these databases considering neXtProt as reference suggests 2830 protein sequences from neXtProt do not have a similar sequence in Ensembl human proteome. Similarly 3896 proteins from neXtProt do not have representative in RefSeq human proteome. On the other hand as many as 13,931 from RefSeq and 47,356 from Ensembl do not have a match in neXtProt database. The scenario further complicates the choice for the database search database even for the most characterized organism like human. It should be noted that most of these differences relate to either splice isoforms or poorly annotated genes. While a better synchronization is required among these primary resources, the differences are primarily due to different genome annotation pipeline adopted by these portals. A similar scenario can be observed for other organisms as well and has been reported even for simple organisms like bacteria. One of the major limitations of the database search method is that a peptide cannot be identified despite its presence in sample, if it is not present in the search database. Therefore, selecting a database with fewer protein entries might lead to underestimation of identified proteins.

2.1.4 Effect of Database Size on Sensitivity and Specificity of Identifications

Primary motivation for searching larger databases is inclusion of most of the biological proteins in search so as to maximize the identifications. However, large databases would also increase the high scoring random matches thus potentially increasing false positives. To control the number of false positives, a decoy database based method is generally adopted to calculate FDR and to filter the identifications. Large target search database would also mean an equally large decoy database and thus high scoring random matches from the decoy database as well. Since number of true target identifications are not expected to increase beyond a finite set with the inflation in database size, the ratio of decoy hits to target hits increases at a given FDR threshold and thus reduces the number of qualifying target identifications. Therefore, rather than maximizing the search results, large database size actually reduces the overall significant identifications as the threshold becomes stringent due to increase in high scoring decoys. A computational correction can be applied where rescorers like Percolator [15] for Mascot, X!Tandem [16] and OMSSA [17], FlexiFDR [18] for MAssWiz [19], and Qscore [20] for Sequest may alleviate this problem to some extent but may be computationally challenging

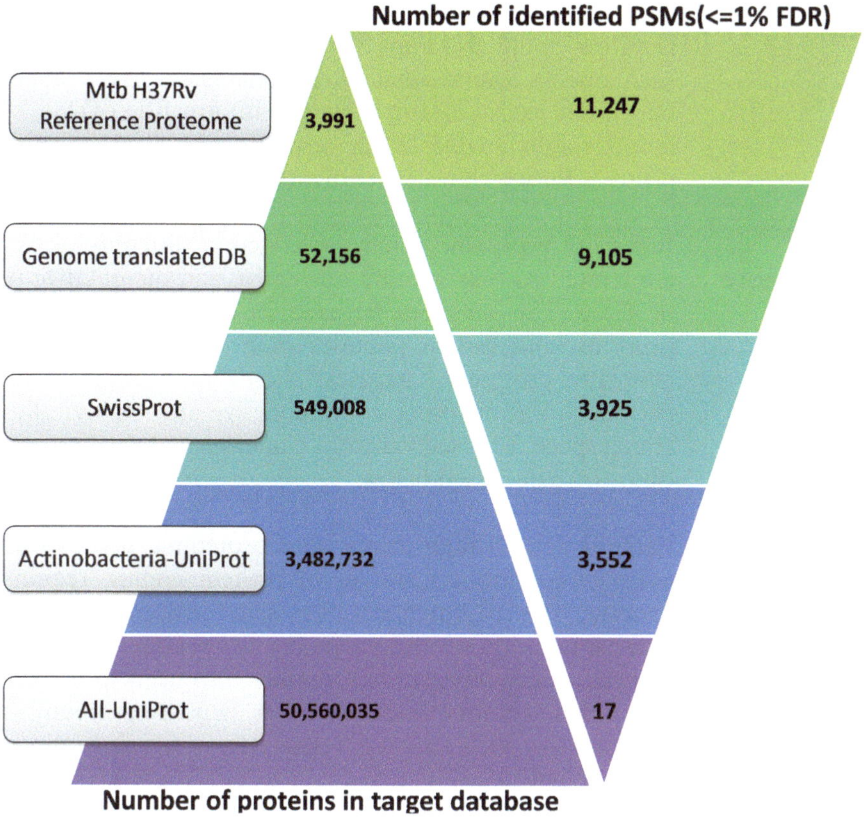

Fig. 3 Size and identified peptide spectrum matches (PSMs) for searches against common databases for Mtb MS/MS data

and outside the domain expertise of general proteomic researchers without bioinformatics support.

To demonstrate this effect, a small dataset from Mtb was searched against five databases of increasing size and complexity; NCBI-Mtb H37Rv reference proteome, six frame translated genome, NCBI proteome for Mtb complex, SwissProt, and UniProtKB. As evident from Fig. 3, increasing database size beyond the expected proteome size tends to decrease the overall number of identifications.

Another aspect of database content is the sharing of peptides among different protein sequences in the database. If search database contains multiple protein isoforms, most of the peptides will be shared among isoforms and clear distinction of expressed isoform(s) is challenging. Protein grouping algorithms are expected to resolve the most probable expressed isoform explaining maximum number of detected peptides [21]. However, its effect on protein quantitation is yet to be characterized. Most of the labeled quantification methods like SILAC [22] and iTRAQ [23], as well as label free ones like SWATH [24], approximate protein quantities from mapping peptide quantities. Hence, peptide sharing among isoforms would also impact the inferred protein quantities. While such analysis against small curated databases like neXtProt might be a useful exercise, sample-specific information of expressed transcripts might enable better resolution in discovering translated proteins for which no experimental evidence exists.

2.2 Custom Databases for Specific Questions

Reference proteome databases hosted at different Web resources are limited to only routine qualitative and quantitative proteome profiling. The questions targeted for discoveries like novel genes, isoforms, and variant peptides cannot be answered using such annotated proteome databases. These require building customized database to cover more comprehensive and biologically relevant search space. Few such custom databases and their targeted applications are discussed below.

2.2.1 Genome Reannotation and Novel Gene Discovery

Large scale and high-throughput proteomic studies can also be used to confirm translation of protein coding regions in any genome. Due to limitation of in silico genome annotation pipelines, annotation of protein coding genes is neither complete nor accurate. Such annotated proteomes thus need to be refined with experimental observations. However, standard proteome databases do not allow such analysis. These *proteogenomic* studies require a database which covers maximal potential coding DNA sequences (CDS) in a genome [25]. A six frame translated genome theoretically includes entire set of CDS. Although 20- to 100-folds larger than annotated proteome databases, genome translated databases have potential to discover genomic regions undergoing translation but not included in annotated set of protein coding genes for the organism [26]. Software developments like GenoSuite [27], EuGenoSuite and Peppy [28] allow proteomic researchers to carry out prokaryotic proteogenomic studies by enabling automated comprehensive data analysis against a genome database. For prokaryotes which have ≈90% coding potential, genome translated databases are preferred choices for proteogenomic reannotation and have been successfully applied to various organisms like Mtb [29], *Bradyrhizobium japonicum* [27], and *Shigella felxneri* [30]. However, for complex eukaryotic organisms like human whose genome has only ≈2% as protein coding regions, analysis against translated genome increases search space many folds and becomes challenging. An alternative is to use ab initio predictions as search database [31], but these depend on the accuracy of the ab initio predictor itself. In such cases translated transcriptome database may serve the purpose.

2.2.2 Detection of Splice Variant Protein Isoforms

Specific to eukaryotes, alternate splicing of messenger RNA expands an organism's protein repertoire excessively. Biogenesis of these protein isoforms is under tight regulation and detection of these might indicate specific biological state and thus hold promise to explain tissue-specific expression patterns. However, most of the proteome databases do not comprise of an exhaustive list of splice isoforms. Recent developments in RNA sequencing (RNA-seq) technologies enable a very deep and sensitive profiling of the transcriptome including even low copy spliced transcripts [32]. A reference database built by theoretically translating the captured transcriptome should thus enable discovery of novel protein isoforms not included in annotated proteomes [33]. RNA-seq also allows tissue-, condition-, or individual-specific transcriptome profiling and thus building sample-specific translated transcriptome database might aid in discovery of rarely detected condition-specific protein isoforms. Proteogenomic analysis pipelines like Enosi [34] and CustomProDB [35] facilitate creation of proteomics search database from raw RNA-seq reads. Several studies have benefitted by discovering isoforms by using RNA-seq based databases to search proteomics data [36–38].

2.2.3 Detection of Polymorphic Peptides

For any species, databases represent only one reference genome and its corresponding reference proteome. However, individuals of the species have differences among their genomes mostly in the form of single nucleotide polymorphisms (SNPs) or insertion/deletion of genomic segments. These genomic differences between individual's genome and the reference genome might also alter the sequence of encoded proteins. Although such variant peptides reflecting nonsynonymous polymorphisms will be present in the biological sample, these cannot be identified by searching against the reference proteome. These polymorphisms are known to be rampant in cancer and their implication in the proteome might be of interest in deciphering the diseases mechanism. For prokaryotes, cumulative database from multiple related strains has been shown to be effective to discover polymorphic peptides [39, 40]. SNP information from resources like dbSNP and COSMIC can also be encoded in protein sequencing to create a thorough database of human protein mutations for peptide detection [41]. Alternatively, sample-specific genome, exome, or transcriptome sequencing might also be utilized to enable sample-specific protein polymorphism detection. Recently several software and packages like CustomProDB have been developed which facilitate creation of variant peptide search database from genomic or RNA-seq based variant calls [35].

2.2.4 Databases for Unsequenced Organisms

Protein identification for organisms whose genome is not sequenced is challenging yet an interesting aspect. It can be especially useful for rare organisms and plants that have limited representation in genome sequencing projects. De novo peptide discovery finds its application in this domain of proteomics, however, with limited sensitivity [42]. Alternate strategy might be to de novo assemble the transcriptome from RNA-Seq data from the unsequenced organism and use it for protein discovery. Brinkman et al. highlighted that this strategy significantly improves peptide discovery over the de novo peptide sequencing in unsequenced box jellyfish [43]. In cases where RNA-seq data are not available, taxonomy information might be helpful. Genome or protein sequences from related organisms might also be utilized as a template to discover proteins from unsequenced organism. However, applicability of such databases is directly proportional to its relatedness to the unsequenced organism. A recently developed algorithm BICEPS promises accurate detection of peptide sequences for an organism by using a database from evolutionary distant organism [44]. This algorithm can bring new proteomic discoveries from unsequenced organisms. Global proteome databases might also be utilized if none of the above mentioned approaches are amenable. A similar strategy might also be useful in community proteomic studies mostly targeted to profile microbiome proteomes from different niches [45].

3 What Makes the Best Proteomics Search Database

Database constitutes arguably the most influencing factor in the protein discovery from tandem mass spectrometry data. Despite its importance, the choice of the database is generally unreasoned. We summarize here salient features of protein databases which should be evaluated before deciding on the search database for a given proteomic data. While organism-specific reference proteomes should be preferred choice over global protein databases, differences among reference proteomes from different sources present a complicated context-specific use case for MS/MS data analysis. A small exercise with only a fraction of overall data to evaluate which reference proteome maximizes the expected results, should facilitate an informed decision on the optimal search database. For targeted studies, like novel gene or isoform discovery, custom database should be designed. However, while designing such dedicated databases, a balance between comprehensiveness and compactness should be maintained.

4 Proteomics Database Resources

1. NCBI-RefSeq: http://www.ncbi.nlm.nih.gov/refseq/.
2. UniProtKB: http://www.uniprot.org/uniprot/.
3. SwissProt: http://www.uniprot.org/uniprot/?query=*&fil=reviewed%3Ayes.

4. UniProt Reference Proteomes: http://www.uniprot.org/proteomes/.
5. neXtProt: http://www.nextprot.org/.
6. Ensembl: http://www.ensembl.org/info/data/ftp/index.html.
7. Tuberculist: http://tuberculist.epfl.ch/.
8. EcoCyc: http://ecocyc.org/.

References

1. Steen H, Mann M (2004) The ABC's (and XYZ's) of peptide sequencing. Nat Rev Mol Cell Biol 5:699–711
2. Fenn JB, Mann M, Meng CK, Wong SF, Whitehouse CM (1989) Electrospray ionization for mass spectrometry of large biomolecules. Science 246:64–71
3. Tanaka K, Waki H, Ido Y, Akita S, Yoshida Y, Yoshida T, Matsuo T (1988) Protein and polymer analyses up to *m/z* 100 000 by laser ionization time-of-flight mass spectrometry. Rapid Commun Mass Spectrom 2:151–153
4. Hunt DF, Yates JR III, Shabanowitz J, Winston S, Hauer CR (1986) Protein sequencing by tandem mass spectrometry. Proc Natl Acad Sci U S A 83:6233–6237
5. Syka JE, Coon JJ, Schroeder MJ, Shabanowitz J, Hunt DF (2004) Peptide and protein sequence analysis by electron transfer dissociation mass spectrometry. Proc Natl Acad Sci U S A 101:9528–9533
6. Dancik V, Addona TA, Clauser KR, Vath JE, Pevzner PA (1999) De novo peptide sequencing via tandem mass spectrometry. J Comput Biol 6:327–342
7. Frank A, Pevzner P (2005) PepNovo: de novo peptide sequencing via probabilistic network modeling. Anal Chem 77:964–973
8. Frank AM, Savitski MM, Nielsen ML, Zubarev RA, Pevzner PA (2007) De novo peptide sequencing and identification with precision mass spectrometry. J Proteome Res 6: 114–123
9. Eng JK, Searle BC, Clauser KR, Tabb DL (2011) A face in the crowd: recognizing peptides through database search. Mol Cell Proteomics 10:R111
10. Kall L, Storey JD, MacCoss MJ, Noble WS (2008) Assigning significance to peptides identified by tandem mass spectrometry using decoy databases. J Proteome Res 7:29–34
11. Elias JE, Gygi SP (2007) Target-decoy search strategy for increased confidence in large-scale protein identifications by mass spectrometry. Nat Methods 4:207–214
12. Gaudet P, Argoud-Puy G, Cusin I, Duek P, Evalet O, Gateau A, Gleizes A, Pereira M, Zahn-Zabal M, Zwahlen C, Bairoch A, Lane L (2013) neXtProt: organizing protein knowledge in the context of human proteome projects. J Proteome Res 12:293–298
13. Harrow J, Frankish A, Gonzalez JM, Tapanari E, Diekhans M, Kokocinski F, Aken BL, Barrell D, Zadissa A, Searle S, Barnes I, Bignell A, Boychenko V, Hunt T, Kay M, Mukherjee G, Rajan J, Despacio-Reyes G, Saunders G, Steward C, Harte R, Lin M, Howald C, Tanzer A, Derrien T, Chrast J, Walters N, Balasubramanian S, Pei B, Tress M, Rodriguez JM, Ezkurdia I, Van BJ, Brent M, Haussler D, Kellis M, Valencia A, Reymond A, Gerstein M, Guigo R, Hubbard TJ (2012) GENCODE: the reference human genome annotation for The ENCODE Project. Genome Res 22:1760–1774
14. Shiromizu T, Adachi J, Watanabe S, Murakami T, Kuga T, Muraoka S, Tomonaga T (2013) Identification of missing proteins in the neXtProt database and unregistered phosphopeptides in the PhosphoSitePlus database as part of the Chromosome-centric Human Proteome Project. J Proteome Res 12:2414–2421
15. Brosch M, Yu L, Hubbard T, Choudhary J (2009) Accurate and sensitive peptide identification with Mascot Percolator. J Proteome Res 8:3176–3181
16. Craig R, Beavis RC (2004) TANDEM: matching proteins with tandem mass spectra. Bioinformatics 20:1466–1467
17. Geer LY, Markey SP, Kowalak JA, Wagner L, Xu M, Maynard DM, Yang X, Shi W, Bryant SH (2004) Open mass spectrometry search algorithm. J Proteome Res 3:958–964
18. Yadav AK, Kumar D, Dash D (2012) Learning from decoys to improve the sensitivity and specificity of proteomics database search results. PLoS One 7, e50651
19. Yadav AK, Kumar D, Dash D (2011) MassWiz: a novel scoring algorithm with target-decoy based analysis pipeline for tandem mass spectrometry. J Proteome Res 10:2154–2160

20. Moore RE, Young MK, Lee TD (2002) Qscore: an algorithm for evaluating SEQUEST database search results. J Am Soc Mass Spectrom 13: 378–386
21. Ma ZQ, Dasari S, Chambers MC, Litton MD, Sobecki SM, Zimmerman LJ, Halvey PJ, Schilling B, Drake PM, Gibson BW, Tabb DL (2009) IDPicker 2.0: improved protein assembly with high discrimination peptide identification filtering. J Proteome Res 8:3872–3881
22. Ong SE, Blagoev B, Kratchmarova I, Kristensen DB, Steen H, Pandey A, Mann M (2002) Stable isotope labeling by amino acids in cell culture, SILAC, as a simple and accurate approach to expression proteomics. Mol Cell Proteomics 1:376–386
23. Ross PL, Huang YN, Marchese JN, Williamson B, Parker K, Hattan S, Khainovski N, Pillai S, Dey S, Daniels S, Purkayastha S, Juhasz P, Martin S, Bartlet-Jones M, He F, Jacobson A, Pappin DJ (2004) Multiplexed protein quantitation in Saccharomyces cerevisiae using amine-reactive isobaric tagging reagents. Mol Cell Proteomics 3:1154–1169
24. Gillet LC, Navarro P, Tate S, Rost H, Selevsek N, Reiter L, Bonner R, Aebersold R (2012) Targeted data extraction of the MS/MS spectra generated by data-independent acquisition: a new concept for consistent and accurate proteome analysis. Mol Cell Proteomics 11:O111
25. Jaffe JD, Berg HC, Church GM (2004) Proteogenomic mapping as a complementary method to perform genome annotation. Proteomics 4:59–77
26. Castellana N, Bafna V (2010) Proteogenomics to discover the full coding content of genomes: a computational perspective. J Proteomics 73: 2124–2135
27. Kumar D, Yadav AK, Kadimi PK, Nagaraj SH, Grimmond SM, Dash D (2013) Proteogenomic analysis of Bradyrhizobium japonicum USDA110 using GenoSuite, an automated multi-algorithmic pipeline. Mol Cell Proteomics 12:3388–3397
28. Risk BA, Spitzer WJ, Giddings MC (2013) Peppy: proteogenomic search software. J Proteome Res 12:3019–3025
29. Kelkar DS, Kumar D, Kumar P, Balakrishnan L, Muthusamy B, Yadav AK, Shrivastava P, Marimuthu A, Anand S, Sundaram H, Kingsbury R, Harsha HC, Nair B, Prasad TS, Chauhan DS, Katoch K, Katoch VM, Kumar P, Chaerkady R, Ramachandran S, Dash D, Pandey A (2011) Proteogenomic analysis of Mycobacterium tuberculosis by high resolution mass spectrometry. Mol Cell Proteomics 10:M111
30. Zhao L, Liu L, Leng W, Wei C, Jin Q (2011) A proteogenomic analysis of Shigella flexneri using 2D LC-MALDI TOF/TOF. BMC Genomics 12:528
31. Ghali F, Krishna R, Perkins S, Collins A, Xia D, Wastling J, Jones AR (2014) ProteoAnnotator – open source proteogenomics annotation software supporting PSI standards. Proteomics 14:2731–2741
32. Wang Z, Gerstein M, Snyder M (2009) RNA-Seq: a revolutionary tool for transcriptomics. Nat Rev Genet 10:57–63
33. Wang X, Liu Q, Zhang B (2014) Leveraging the complementary nature of RNA-Seq and shotgun proteomics data. Proteomics 14:2676–2687
34. Castellana NE, Shen Z, He Y, Walley JW, Cassidy CJ, Briggs SP, Bafna V (2014) An automated proteogenomic method uses mass spectrometry to reveal novel genes in Zea mays. Mol Cell Proteomics 13:157–167
35. Wang X, Zhang B (2013) CustomProDB: an R package to generate customized protein databases from RNA-Seq data for proteomics search. Bioinformatics 29:3235–3237
36. Sun H, Chen C, Shi M, Wang D, Liu M, Li D, Yang P, Li Y, Xie L (2014) Integration of mass spectrometry and RNA-Seq data to confirm human ab initio predicted genes and lncRNAs. Proteomics 14:2760–2768
37. Woo S, Cha SW, Merrihew G, He Y, Castellana N, Guest C, MacCoss M, Bafna V (2014) Proteogenomic database construction driven from large scale RNA-seq data. J Proteome Res 13:21–28
38. Omasits U, Quebatte M, Stekhoven DJ, Fortes C, Roschitzki B, Robinson MD, Dehio C, Ahrens CH (2013) Directed shotgun proteomics guided by saturated RNA-seq identifies a complete expressed prokaryotic proteome. Genome Res 23:1916–1927
39. de Souza GA, Arntzen MO, Wiker HG (2010) MSMSpdbb: providing protein databases of closely related organisms to improve proteomic characterization of prokaryotic microbes. Bioinformatics 26:698–699
40. de Souza GA, Arntzen MO, Fortuin S, Schurch AC, Malen H, McEvoy CR, Van SD, Thiede B, Warren RM, Wiker HG (2011) Proteogenomic analysis of polymorphisms and gene annotation divergences in prokaryotes using a clustered mass spectrometry-friendly database. Mol Cell Proteomics 10:M110
41. Nagaraj SH, Waddell N, Madugundu AK, Wood S, Jones A, Mandyam RA, Nones K, Pearson JV, Grimmond SM (2015) PGTools: a software suite for proteogenomic data analysis and visualization. J Proteome Res 14:2255–2266

42. Brinkman DL, Aziz A, Loukas A, Potriquet J, Seymour J, Mulvenna J (2012) Venom proteome of the box jellyfish Chironex fleckeri. PLoS One 7, e47866
43. Brinkman DL, Jia X, Potriquet J, Kumar D, Dash D, Kvaskoff D, Mulvenna J (2015) Transcriptome and venom proteome of the box jellyfish Chironex fleckeri. BMC Genomics 16:407
44. Renard BY, Xu B, Kirchner M, Zickmann F, Winter D, Korten S, Brattig NW, Tzur A, Hamprecht FA, Steen H (2012) Overcoming species boundaries in peptide identification with Bayesian information criterion-driven error-tolerant peptide search (BICEPS). Mol Cell Proteomics 11:M111
45. Delmotte N, Knief C, Chaffron S, Innerebner G, Roschitzki B, Schlapbach R, Von MC, Vorholt JA (2009) Community proteogenomics reveals insights into the physiology of phyllosphere bacteria. Proc Natl Acad Sci U S A 106:16428–16433

Chapter 4

Label-Based and Label-Free Strategies for Protein Quantitation

Sushma Anand*, Monisha Samuel*, Ching-Seng Ang, Shivakumar Keerthikumar, and Suresh Mathivanan

Abstract

The precise quantification of changes between various physiological states in a biological system is highly complex in nature. Over the past few years, in combination with classical methods, mass spectrometry based approaches have become an indispensable tool in deciphering exact abundance of proteins in composite mixtures. The technique is now well established and employs both label-based and label-free quantitation strategies. Label-based quantitation methods utilize stable isotope labels which are incorporated within the peptides, introducing an expectable mass difference within the two or more experimental conditions. In contrast, label-free proteomics quantitates both relative and absolute protein quantity by utilizing signal intensity and spectral counting of peptides. This chapter focuses on the commonly used quantitative mass spectrometry methods for high-throughput proteomic analysis.

Key words Quantitative proteomics, Mass spectrometry, Stable isotope labeling, Label-free quantitation

1 Introduction

Unlike the finite genome, the proteome is perturbed both temporally and spatially within a cell [1]. The complexity is multifaceted depending upon the epigenetic status, posttranscriptional events, posttranslational events, and physiological stimuli [2, 3]. Hence, functional and quantitative characterization of every human protein based on their isoforms, posttranslational modifications, subcellular localization, tissue expression, and protein interaction partners is indeed a daunting task. Recent advances in mass spectrometry-based proteomics techniques have allowed for the production of a draft map of the human proteome [4, 5], a decade after the completion of the human genome project [6, 7]. Using an array of supplemental techniques, identification and

*Authors contributed equally with all other contributors.

Shivakumar Keerthikumar and Suresh Mathivanan (eds.), *Proteome Bioinformatics*, Methods in Molecular Biology, vol. 1549, DOI 10.1007/978-1-4939-6740-7_4,

quantification of a large number of proteins in a high-throughput manner can be carried out [8, 9]. Using existing functional enrichment analysis tools, the obtained proteomic data can be analyzed in the context of biological pathways and protein interaction networks to understand their contribution to pathophysiology [10].

Mass spectrometry-based quantitative proteomics can be classified into two broad categories—label-based and label-free methods. Label-based methods in proteomics have been made available through the introduction of ICAT labels by the Aebersold's group [11]. Since then, there have been a large variety of modifications and adaptations to that technique. Essentially, quantitation of proteins is based on light/heavy peptide intensities. In label-based methods, samples are first differentially labeled, pooled, and subjected to MS analysis and quantification. Hence, it minimizes the disparities expected when samples are handled individually [12, 13]. The most widely used labeling techniques include metabolic, proteolytic, and chemical labeling strategies. Recently, there is significant interest in label-free quantitative proteomics, resulting from the introduction of high resolution/accurate mass spectrometers and also its ease of use and reproducibility. Here, the samples are processed and analyzed independently by mass spectrometry. The ensuing quantification is performed by the measurement of the peak area and/or consideration of the number of MS/MS spectra from each peptide [12, 14]. To aid in quantitation, various software programs are used to analyze the large amount of data generated using both label-based and label-free techniques (Table 1). Throughout this chapter, emphasis is placed on MS-based protein quantification using label-based and label-free strategies with perks and pitfalls.

2 Methods

2.1 Label-Based Quantitation Strategies

Label-based quantitation involves comparison of samples by labeling them with alternative differential mass tags thus allowing detection based on specific change in mass. This is a comparative approach which in general employs incorporation of chemically similar but isotopically different labels [15]. Thus the labeling strategy easily divulges both relative as well as absolute quantitation of proteins from individual samples within the same run.

2.1.1 Stable Isotope Introduction by Metabolic Labeling

Metabolic labeling is one of the most preferred methods of labeling as it allows least experimental variability due to introduction of labels at the earliest possible stages of sample preparation [12]. In this process, isotopically defined medium is used to introduce distinct labels into the proteome of two or more biologically different samples during the process of protein metabolism. The samples are then equally pooled and analyzed [16]. This creates two versions of each peptide with different isotopic compositions but with

Table 1
Common software packages available for analysis of quantitative proteomics data

Label-based quantitation	SILAC	BioWorks, Census, Mascot Distiller, Elucidator, MaxQuant, MaXIC-Q, OpenMS, PeakQuant, MFPaQ, PEAKS Q, ProteinPilot, ProteoIQ, TPP-ASAPRatio, WARP-LC, MSQuant
	^{15}N, ^{18}O	^{18}O: Mascot Distiller, MSQuant, PEAKS Q, ProteoIQ, QUIL, STEM, VIPER, ZoomQuant, ProRata ^{15}N: MSQuant, Census, PeakQuant, ProRata, ProteoIQ, Qupe, TPP-XPRESS, X-TRacker
	ICAT, ITRAQ, TMT	ICAT: BioWorks, Elucidator, MaxQuant, MaXIC-Q, MFPaQ, PEAKS Q, ProteinPilot, ProteoIQ, TPP-ASAPRatio, MSQuant, QUIL, TPP-XPRESS, VIPER, ProRata ITRAQ: BioWorks, Census, ITracker, OPenMS, PeakQuant, PEAKS Q, Pro Quant, ProteinPilot, Proteios, ProteoIQ, Proteome Discoverer, TPP-Libra, X-Tracker TMT: Proteios, ProteoIQ, Proteome Discover, PEAKS Q
Label-free quantitation	Based on peak intensity	SpecArray, MSight, PEPPeR, MSInspect, MSQuant, Census, Corra, Serac, SuerHIrn, MzMine, BioWorks, Elucidator, Mascot Distiller, OpenMS, ProteoIQ, SIEVE, Skyline
	Based on spectral count	SEQUEST, MASCOT, X!Tandem, ProteoIQ, Census, PepC, emPAI Calc, Elucidator, MFPaQ

identical response to sample preparation and MS analysis, thus allowing relative quantitation of the labeled sample without alteration of their biochemical properties [17]. Initially, this began with the usage of ^{15}N substituted media, in which all the ^{14}N atoms were replaced by ^{15}N, leading to differential quantification between the states of microorganisms [18]. Various snags including partial incorporation of labels, higher expenses and difficulties in data analysis have rendered this method less advantageous [16].

A more efficient technique known as SILAC (Stable isotope labeling of amino acids in cell culture) was then introduced by Mann et al. It is a nonselective method of labeling proteins in vivo, where heavy forms of essential amino acids are generally used as labels. It involves usage of different medium for growing two populations of cells, one containing light (normal) amino acids and the other containing heavy (isotopically labeled) amino acids. This labeling is obtained by replacing the naturally occurring elements of H, ^{14}N and ^{12}C to ^{2}H, ^{15}N, and ^{13}C, respectively. The samples are further mixed, fractionated, and analyzed by MS. The labeled amino acids thus gets incorporated into all the newly synthesized proteins and hence are encoded into the proteome [19]. The analysis further distinguishes two proteomes by the molecular weight of the light and heavy amino acids which was used during the growth of the two cell population [20]. The commonly chosen amino acids to achieve effective labeling are leucine, lysine, methionine, and arginine [16]. SILAC offers higher incorporation rate, thus enhancing labeling efficiency, not requiring chemical manipulation, and also reducing sample handling error as the labels are mixed in the very initial stages. When used with well-designed combination of labeled amino acids, SILAC offers the ability to compare as many as five states within a single experiment. It has already been widely used to study posttranslational modifications, enzyme substrates studies, identification of cancer biomarkers, understanding protein complexes and signaling pathways [19]. SILAC has its own drawbacks and one of it is that the number of cellular states that can be compared becomes restricted due to the constraint in availability of various ranges of heavy forms of amino acids [21]. Despite a few shortcomings, SILAC quantitative labeling method offers several major benefits and is one of the most effective labeling strategies used for quantitative proteomics.

2.1.2 Stable Isotope Introduction by Chemical Labeling

Chemical labeling is one of the most frequently used method in proteomics research as it offers the flexibility of selective introduction of isotopic labels into desired position in a peptide or a protein. It has a similar principle as metabolic labeling except that chemical reactions are used to incorporate these labels into their desired locations. Here, we broadly describe several of the most widely used chemical labeling approaches.

Isotope-coded affinity tagging (ICAT) was one of the earliest methods introduced by Aebersold et al. [11], which was employed to study the yeast proteome. ICAT reagent contains an iodoacetyl reactive group to target cysteinyl residues, a linker region consisting heavy ($^{2}H_{8}$) or light ($^{1}H_{8}$) deuterium atoms and a biotin group suitable for affinity purification [11]. The thiol chemistry labels only the cysteine residues of the proteins whereas the biotin group aids in the capture of cysteine-containing peptides specifically based on biotin–avidin affinity. This greatly reduces sample complexity thus simplifying ensuing MS analysis and subsequent interpretation. One of the major pitfalls of this method was the shift in retention time in the light and heavy peptides during reverse-phase chromatography caused by the deuterium atoms present in the linker region [22]. Therefore, recently, newer variants of ICAT have been introduced like cleavable ICAT (cICAT) where acid-cleavable isotopic tags are used with ^{13}C, $^{13}C_{9}$, or $^{12}C_{9}$ in place of deuterium atoms which improves peptide recovery and ease of automation [23].

Dimethyl labeling was introduced as a technique in quantitative proteomics in the year 2003 by Hsu et al. [23]. The basics behind dimethyl labeling involves the formation of a Schiff base via the reaction of formaldehyde with primary amines in a near-neutral pH, which is further reduced by cyanoborohydride and formaldehyde. The usage of different isotopomers of formaldehyde in combination with cyanoborohydride generates difference in mass per labeling event [24]. The strength of this labeling method lies in its inexpensive reagents, quicker reaction mechanism without generation of any significant side products as well as wider applicability to many types of sample. One of the drawbacks frequently associated with dimethyl labeling is the overlap of retention time shifts between the heavy labeled peptides and their lighter counterparts during reverse phase chromatography. Recently, strategies have also been devised to overcome this issue [25].

Isobaric labeling methods like tandem mass tag (TMT) and isobaric tag for absolute and relative quantitation (iTRAQ) are other forms of chemical labeling which are amine-specific in nature. The proteins are thus labeled with chemical groups which are isobaric (identical in mass) in nature but dissociate under tandem MS to yield reporter ions of variable mass. Like in case of iTRAQ, all primary amine functional groups of the peptides are tagged with a peptide reactive group, a balancer, and a reporter group. As the technique targets the amine group, almost all peptides present in the sample are labeled and quantified easily. The reporter group contains 4–8 different tagged sites, and this creates a mass difference which is counteracted by the balancer group, thus making the peptides similar in terms of mass [16]. After labeling, MS/MS is used to fragment the iTRAQ reagents to generate reporter ions in a distinct mass range (113–121 Da), and the amount of reporter ion released is directly proportional to the amount of the tagged peptides in the samples under comparison [26]. TMT is designed to multiplex a maximum of ten samples whereas iTRAQ provides the scope of analyzing up to eight different samples in a single run [16]. The application of iTRAQ has become widespread as it can be used simultaneously for multiple samples and offers a deeper coverage with higher quantification precision. However, there are several drawbacks as well like higher cost, nonspecificity of reporter ions for various peptides, impurities associated with the reporter ions, requirement of quad based instrument with high resolving powers, and issues associated with co-isolation of peptides during precursor selection [27].

2.1.3 Stable Isotope Introduction by Enzymatic Labeling

^{18}O labeling is a technique which employs proteolytic catalysis by class-2 proteases such as trypsin to incorporate two ^{18}O atoms in the place of ^{16}O atoms, resulting in a mass shift between the differently labeled peptides [28]. This is a two-step mechanism with an initial hydrolysis reaction which is followed by a protease-aided incorporation of ^{18}O atom into the carboxyl terminus of the proteolytically generated peptide. After performing the digestion as well as labeling simultaneously, the two sets of samples are further pooled for sample preparation and MS analysis. Enzymes like trypsin, chymotrypsin, and Glu-C have been of primary choice for this method [16]. Despite its simplicity, the technique has not been widely used in quantitative proteomics due to its various pitfalls. When both the samples are mixed together for further processing, back exchange can occur as the enzyme can still act on the ^{16}O labeled c-terminus and this can alter the $^{16}O/^{18}O$ ratio. This, however, can be overcome by using immobilized trypsin and using quantitation algorithms [29]. Partial incorporation of the labels is also an issue that has to be addressed to improve the efficiency as well as utility of this technique [30].

2.2 Label-Free Quantitation Strategies

Label-free quantitative proteomics (LFQP) provides straightforward option for large-scale analysis of biological samples. In contrast to label-based methods, samples of interest to be compared are injected independently into MS [14, 31]. LFQP has several advantages over label-based quantitative proteomics as it is cost-effective and does not require expensive labeling reagents. Also, LFQP is not time consuming as compared to some of the label-based methods which requires tedious labeling steps [32]. Due to all these aforementioned reasons, LFQP has gained more acceptance in biomedical research space. LFQP is a very powerful technique, which is less susceptible to technical error and highly sensitive to MS analysis thus enabling identification of several thousand proteins from complex samples such as bodily fluids (blood, plasma, saliva, and urine), cell lines, and tissues [33–35].

Outmoded quantitative proteomics involves two-dimensional gel electrophoresis coupled (2D-GE) with liquid mass spectrometry (LC-MS). This procedure includes the separation of proteins based on their isoelectric point in the first dimension followed by separation on the basis of molecular weight using sodium dodecyl sulfate polyacrylamide gel electrophoresis (SDS-PAGE) followed by staining of gel with protein specific dyes and fluorophores. The protein spots of interest are then digested and peptides are extracted followed by mass spectrometry analysis. Quantitation is carried out by comparing the protein spot intensities across multiple gel replicates. Although this method is quite sensitive and precise, it requires large amount of sample to be processed along with various sample preparation steps [35, 36]. Due to these downsides and recent advancements in mass spectrometry based chromatographic methods, various quantitative mass spectrometry approaches are being more frequently used for high-throughput and large-scale protein analysis [35].

Several label-free quantitation methods are developed for the quantification of proteins identified using tandem mass spectrometry. Here, we discuss the most commonly used label-free quantitation methods based on spectral counting and peak intensity for comparative analysis of the relative abundance of proteins. A brief overview on workflow of label-free quantitation is described in Fig. 1.

2.2.1 Spectral Counting (SpC)

Spectral counting (SpC) is one of the most widely used label-free quantitative methods in the field of proteomic analysis. The principle of SpC relies on frequency and abundance of the protein. Hence, size and amount of a protein present in the sample is

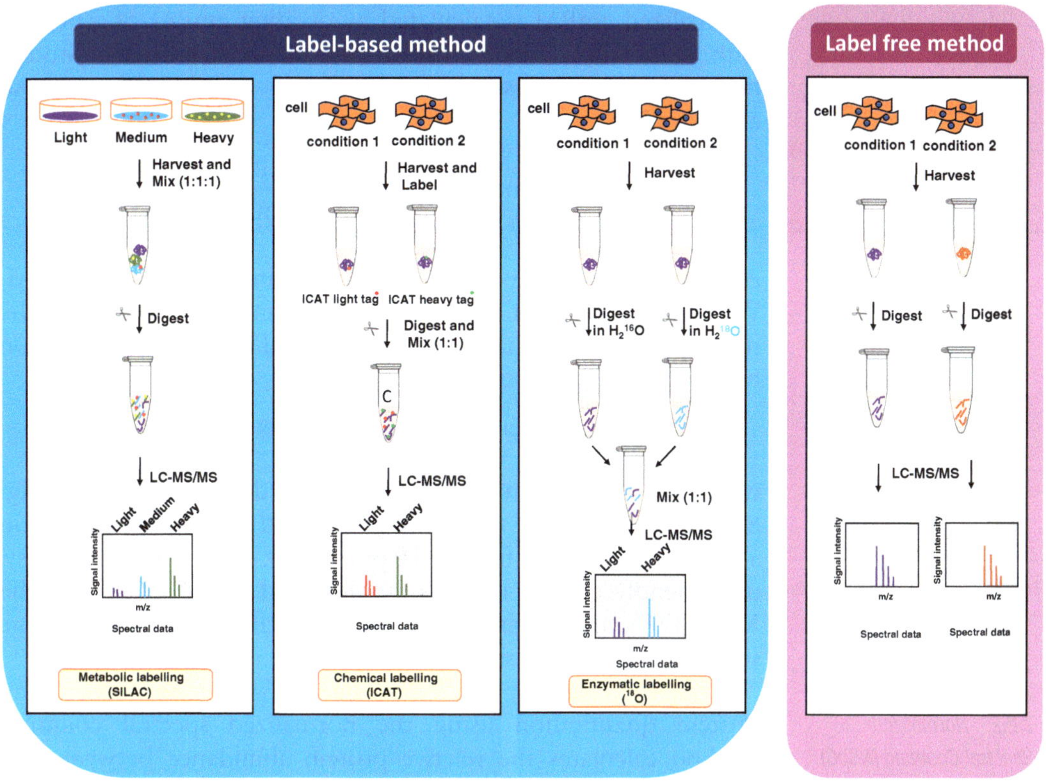

Fig 1 General scheme of quantitative proteomics. (**a**) Label-based approach includes the incorporation of the stable isotope label at different stages depending on the type of technique employed and simultaneous profiling of more than one biological sample in a single MS run. The labeled samples are mixed in an equivalent ratio and analyzed by LC-MS/MS. The quantitation is based on the comparison of peak intensity ratio of the labeled peptide pairs. (**b**) In label-free method, the samples are separately prepared and are subjected to individual LC-MS/MS analysis. Further quantification is based on comparing counts of MS/MS spectra or peak intensity of the same peptide

directly proportional to number of identified spectra. Therefore for comparative estimation, spectral counts of proteins from two independent samples are equated.

Several spectral counting methodologies have been reported in the literature. Protein abundance index (PAI) is an SpC quantification method to calculate the abundance of proteins in a given sample [37]. PAI is defined as the number of observed peptides divided by the number of theoretically observable tryptic peptides for each protein within the mass range of the used MS instrument.

$$\text{PAI} = \frac{\text{Number of observed peptides}}{\text{Number of observable tryptic peptides (theoretical)}}.$$

The representation of PAI was later modified to exponentially modified protein abundance index also referred to as emPAI [38, 39].

$$\text{emPAI} = 10^{\text{PAI}} - 1.$$

Normalized emPAI is calculated by dividing emPAI of a protein with summation of all emPAI of identified proteins. Studies have shown closer association of emPAI with absolute protein amount [38].

$$\text{Normalized emPAI} = \frac{\text{emPAI}}{\sum \text{emPAI}}.$$

Absolute protein amount is calculated using normalized emPAI. When the amount of total protein of a sample processed for the mass spectrometry analysis is known, total protein concentration can be equated as below:

$$\text{Protein concentration} = \text{Normalized emPAI} \times C.$$

where C is the known protein amount used in sample processing. Studies have shown that emPAI method can quantify protein abundance within 10 fmol to 10 pmol [40]. emPAI method is easy, user friendly and its outputs are accepted and displayed by the MASCOT search engine, and thus it is specifically suitable for large scale proteomic analysis [37]

2.2.2 Normalized Spectral Counting (NSpC)

Protein quantitation using the normalized spectral counting method calculates the relative protein abundance between two samples [41–43]. It accounts for the ratio of the total number of spectra identified for each protein of interest normalized by unique spectra identified in the protein samples.

$$\text{RSc for protein A} = \left[(sY + c)(TX - sX + c) / (sX + c)(TY - sY + c)\right]$$

Where RSc represent ratio of normalized spectral counts, s is the significant MS/MS spectra for protein A, T is the total number of significant MS/MS spectra in the sample, c is the correction factor set to 1.25, and X and Y are the two different samples. When RSc is less than 1, the negative inverse of RSc value is used.

2.2.3 Normalized Spectral Abundance Factor (NSAF)

A further refinement in the quantification of the proteins using label-free method was developed which takes into account sample to sample variation and also the notion that longer proteins tends to be identified with more peptides in comparison to shorter proteins [44, 45]. NSAF calculates number of spectral counts (SpC) of a protein divided by its length (L) and normalized to the total sum of spectral counts (SpC)/length (L) of all proteins in a given analysis.

$$\text{NSAF} = \text{SAF} / \sum_{i=1}^{N} \text{SAF},$$

where spectral abundance factor (SAF) represents SpC/L and N represent total number of proteins identified.

2.2.4 Area Under Curve

Area under the curve measures and compares signal intensities of chromatograms obtained by MS for specific peptide. Precursor ion chromatogram for each peptide is extracted from individual LC-MS/MS run and their peak region (area under the curve (AUC)) acquired by MS are integrated over the retention time [12, 36]. It must be noted that for accuracy in protein estimation, individual aligned peak should correspond to its precursor ion, retention time, charge status, and fragmented ion. After accurate alignment process, identified peptides from each samples are measured for AUC and then equated for comparative protein amount [34]. Neilson et al. have described that the measurement of AUC is relative to the abundance of peptide in a specific sample [37]. One of the drawbacks of peak intensity quantification is co-elution of peptides due to its spread over retention time which causes difficulty in identification of two individual peptides. In addition, biological distinctions cause variations in elution time, MS signals, and background noise, and thus it is necessary to filter the raw MS data [37]. Additionally, peptide quantification by AUC depends significantly on raw MS data analysis by software. AUC data quantification also includes data normalization which helps in exclusion of the background noise obtained during multiple MS analysis, thus minimizing the undesirable systematic errors generated during sample preparation [34].

3 Conclusion

Numerous methods have emerged over the years for the analysis of various proteomes using quantitative proteomics. With the fast-paced development of sensitive MS instruments, the usage of quantitative MS has hugely aided in distinguishing perturbations in protein expression and associated changes with posttranslational modifications and protein–protein interactions. This has led to a better understanding of the underlying biology and effective designing of follow-up experiments. Both labeling and label-free approaches have their own sets of strengths and limitations. The method of choice depends to a large extent on the biological question, the researcher, the cost involved as well as the quality of the available MS instrument. At the same time, significant development of various bioinformatics and statistical tools is absolutely necessary to simplify the complexity as well as manage the sheer volume of data generated using these techniques. This can only be achieved by analyzing the implications on the results when these various strategies are employed.

References

1. Mathivanan S (2014) Integrated bioinformatics analysis of the publicly available protein data shows evidence for 96% of the human proteome. J Proteomics Bioinform 7:41–49
2. Kuster B, Schirle M, Mallick P, Aebersold R (2005) Scoring proteomes with proteotypic peptide probes. Nat Rev Mol Cell Biol 6(7): 577–583
3. Nilsen TW, Graveley BR (2010) Expansion of the eukaryotic proteome by alternative splicing. Nature 463(7280):457–463
4. Kim MS, Pinto SM, Getnet D, Nirujogi RS, Manda SS, Chaerkady R, Madugundu AK, Kelkar DS, Isserlin R, Jain S, Thomas JK, Muthusamy B, Leal-Rojas P, Kumar P, Sahasrabuddhe NA, Balakrishnan L, Advani J, George B, Renuse S, Selvan LD, Patil AH, Nanjappa V, Radhakrishnan A, Prasad S, Subbannayya T, Raju R, Kumar M, Sreenivasamurthy SK, Marimuthu A, Sathe GJ, Chavan S, Datta KK, Subbannayya Y, Sahu A, Yelamanchi SD, Jayaram S, Rajagopalan P, Sharma J, Murthy KR, Syed N, Goel R, Khan AA, Ahmad S, Dey G, Mudgal K, Chatterjee A, Huang TC, Zhong J, Wu X, Shaw PG, Freed D, Zahari MS, Mukherjee KK, Shankar S, Mahadevan A, Lam H, Mitchell CJ, Shankar SK, Satishchandra P, Schroeder JT, Sirdeshmukh R, Maitra A, Leach SD, Drake CG, Halushka MK, Prasad TS, Hruban RH, Kerr CL, Bader GD, Iacobuzio-Donahue CA, Gowda H, Pandey A (2014) A draft map of the human proteome. Nature 509(7502):575–581. doi:10.1038/nature13302
5. Wilhelm M, Schlegl J, Hahne H, Moghaddas Gholami A, Lieberenz M, Savitski MM, Ziegler E, Butzmann L, Gessulat S, Marx H, Mathieson T, Lemeer S, Schnatbaum K, Reimer U, Wenschuh H, Mollenhauer M, Slotta-Huspenina J, Boese JH, Bantscheff M, Gerstmair A, Faerber F, Kuster B (2014) Mass-spectrometry-based draft of the human proteome. Nature 509(7502):582–587. doi:10.1038/nature13319
6. Venter JC, Adams MD, Myers EW, Li PW, Mural RJ, Sutton GG, Smith HO, Yandell M, Evans CA, Holt RA, Gocayne JD, Amanatides P, Ballew RM, Huson DH, Wortman JR, Zhang Q, Kodira CD, Zheng XH, Chen L, Skupski M, Subramanian G, Thomas PD, Zhang J, Gabor Miklos GL, Nelson C, Broder S, Clark AG, Nadeau J, McKusick VA, Zinder N, Levine AJ, Roberts RJ, Simon M, Slayman C, Hunkapiller M, Bolanos R, Delcher A, Dew I, Fasulo D, Flanigan M, Florea L, Halpern A, Hannenhalli S, Kravitz S, Levy S, Mobarry C, Reinert K, Remington K, Abu-Threideh J, Beasley E, Biddick K, Bonazzi V, Brandon R, Cargill M, Chandramouliswaran I, Charlab R, Chaturvedi K, Deng Z, Di Francesco V, Dunn P, Eilbeck K, Evangelista C, Gabrielian AE, Gan W, Ge W, Gong F, Gu Z, Guan P, Heiman TJ, Higgins ME, Ji RR, Ke Z, Ketchum KA, Lai Z, Lei Y, Li Z, Li J, Liang Y, Lin X, Lu F, Merkulov GV, Milshina N, Moore HM, Naik AK, Narayan VA, Neelam B, Nusskern D, Rusch DB, Salzberg S, Shao W, Shue B, Sun J, Wang Z, Wang A, Wang X, Wang J, Wei M, Wides R, Xiao C, Yan C, Yao A, Ye J, Zhan M, Zhang W, Zhang H, Zhao Q, Zheng L, Zhong F, Zhong W, Zhu S, Zhao S, Gilbert D, Baumhueter S, Spier G, Carter C, Cravchik A, Woodage T, Ali F, An H, Awe A, Baldwin D, Baden H, Barnstead M, Barrow I, Beeson K, Busam D, Carver A, Center A, Cheng ML, Curry L, Danaher S, Davenport L, Desilets R, Dietz S, Dodson K, Doup L, Ferriera S, Garg N, Gluecksmann A, Hart B, Haynes J, Haynes C, Heiner C, Hladun S, Hostin D, Houck J, Howland T, Ibegwam C, Johnson J, Kalush F, Kline L, Koduru S, Love A, Mann F, May D, McCawley S, McIntosh T, McMullen I, Moy M, Moy L, Murphy B, Nelson K, Pfannkoch C, Pratts E, Puri V, Qureshi H, Reardon M, Rodriguez R, Rogers YH, Romblad D, Ruhfel B, Scott R, Sitter C, Smallwood M, Stewart E, Strong R, Suh E, Thomas R, Tint NN, Tse S, Vech C, Wang G, Wetter J, Williams S, Williams M, Windsor S, Winn-Deen E, Wolfe K, Zaveri J, Zaveri K, Abril JF, Guigo R, Campbell MJ, Sjolander KV, Karlak B, Kejariwal A, Mi H, Lazareva B, Hatton T, Narechania A, Diemer K, Muruganujan A, Guo N, Sato S, Bafna V, Istrail S, Lippert R, Schwartz R, Walenz B, Yooseph S, Allen D, Basu A, Baxendale J, Blick L, Caminha M, Carnes-Stine J, Caulk P, Chiang YH, Coyne M, Dahlke C, Mays A, Dombroski M, Donnelly M, Ely D, Esparham S, Fosler C, Gire H, Glanowski S, Glasser K, Glodek A, Gorokhov M, Graham K, Gropman B, Harris M, Heil J, Henderson S, Hoover J, Jennings D, Jordan C, Jordan J, Kasha J, Kagan L, Kraft C, Levitsky A, Lewis M, Liu X, Lopez J, Ma D, Majoros W, McDaniel J, Murphy S, Newman M, Nguyen T, Nguyen N, Nodell M, Pan S, Peck J, Peterson M, Rowe W, Sanders R, Scott J, Simpson M, Smith T, Sprague A, Stockwell T, Turner R, Venter E, Wang M, Wen M, Wu D, Wu M, Xia A, Zandieh A, Zhu X (2001) The sequence of the human genome. Science 291(5507):1304–1351
7. Lander ES, Linton LM, Birren B, Nusbaum C, Zody MC, Baldwin J, Devon K, Dewar K, Doyle M, FitzHugh W, Funke R, Gage D, Harris K, Heaford A, Howland J, Kann L, Lehoczky J, LeVine R, McEwan P, McKernan K, Meldrim J, Mesirov JP, Miranda C, Morris

W, Naylor J, Raymond C, Rosetti M, Santos R, Sheridan A, Sougnez C, Stange-Thomann N, Stojanovic N, Subramanian A, Wyman D, Rogers J, Sulston J, Ainscough R, Beck S, Bentley D, Burton J, Clee C, Carter N, Coulson A, Deadman R, Deloukas P, Dunham A, Dunham I, Durbin R, French L, Grafham D, Gregory S, Hubbard T, Humphray S, Hunt A, Jones M, Lloyd C, McMurray A, Matthews L, Mercer S, Milne S, Mullikin JC, Mungall A, Plumb R, Ross M, Shownkeen R, Sims S, Waterston RH, Wilson RK, Hillier LW, McPherson JD, Marra MA, Mardis ER, Fulton LA, Chinwalla AT, Pepin KH, Gish WR, Chissoe SL, Wendl MC, Delehaunty KD, Miner TL, Delehaunty A, Kramer JB, Cook LL, Fulton RS, Johnson DL, Minx PJ, Clifton SW, Hawkins T, Branscomb E, Predki P, Richardson P, Wenning S, Slezak T, Doggett N, Cheng JF, Olsen A, Lucas S, Elkin C, Uberbacher E, Frazier M, Gibbs RA, Muzny DM, Scherer SE, Bouck JB, Sodergren EJ, Worley KC, Rives CM, Gorrell JH, Metzker ML, Naylor SL, Kucherlapati RS, Nelson DL, Weinstock GM, Sakaki Y, Fujiyama A, Hattori M, Yada T, Toyoda A, Itoh T, Kawagoe C, Watanabe H, Totoki Y, Taylor T, Weissenbach J, Heilig R, Saurin W, Artiguenave F, Brottier P, Bruls T, Pelletier E, Robert C, Wincker P, Smith DR, Doucette-Stamm L, Rubenfield M, Weinstock K, Lee HM, Dubois J, Rosenthal A, Platzer M, Nyakatura G, Taudien S, Rump A, Yang H, Yu J, Wang J, Huang G, Gu J, Hood L, Rowen L, Madan A, Qin S, Davis RW, Federspiel NA, Abola AP, Proctor MJ, Myers RM, Schmutz J, Dickson M, Grimwood J, Cox DR, Olson MV, Kaul R, Raymond C, Shimizu N, Kawasaki K, Minoshima S, Evans GA, Athanasiou M, Schultz R, Roe BA, Chen F, Pan H, Ramser J, Lehrach H, Reinhardt R, McCombie WR, de la Bastide M, Dedhia N, Blocker H, Hornischer K, Nordsiek G, Agarwala R, Aravind L, Bailey JA, Bateman A, Batzoglou S, Birney E, Bork P, Brown DG, Burge CB, Cerutti L, Chen HC, Church D, Clamp M, Copley RR, Doerks T, Eddy SR, Eichler EE, Furey TS, Galagan J, Gilbert JG, Harmon C, Hayashizaki Y, Haussler D, Hermjakob H, Hokamp K, Jang W, Johnson LS, Jones TA, Kasif S, Kaspryzk A, Kennedy S, Kent WJ, Kitts P, Koonin EV, Korf I, Kulp D, Lancet D, Lowe TM, McLysaght A, Mikkelsen T, Moran JV, Mulder N, Pollara VJ, Ponting CP, Schuler G, Schultz J, Slater G, Smit AF, Stupka E, Szustakowski J, Thierry-Mieg D, Thierry-Mieg J, Wagner L, Wallis J, Wheeler R, Williams A, Wolf YI, Wolfe KH, Yang SP, Yeh RF, Collins F, Guyer MS, Peterson J, Felsenfeld A, Wetterstrand KA, Patrinos A, Morgan MJ, de Jong P, Catanese JJ, Osoegawa K, Shizuya H, Choi S, Chen YJ (2001) Initial sequencing and analysis of the human genome. Nature 409(6822):860–921

8. Boja ES, Rodriguez H (2012) Mass spectrometry-based targeted quantitative proteomics: achieving sensitive and reproducible detection of proteins. Proteomics 12(8):1093–1110. doi:10.1002/pmic.201100387
9. Chisanga D, Keerthikumar S, Pathan M, Ariyaratne D, Kalra H, Boukouris S, Mathew NA, Saffar HA, Gangoda L, Ang CS, Sieber OM, Mariadason JM, Dasgupta R, Chilamkurti N, Mathivanan S (2016) Colorectal cancer atlas: an integrative resource for genomic and proteomic annotations from colorectal cancer cell lines and tissues. Nucleic Acids Res 44(D1): D969–D974. doi:10.1093/nar/gkv1097
10. Pathan M, Keerthikumar S, Ang CS, Gangoda L, Quek CY, Williamson NA, Mouradov D, Sieber OM, Simpson RJ, Salim A, Bacic A, Hill AF, Stroud DA, Ryan MT, Agbinya JI, Mariadason JM, Burgess AW, Mathivanan S (2015) FunRich: an open access standalone functional enrichment and interaction network analysis tool. Proteomics 15(15):2597–2601. doi:10.1002/pmic.201400515
11. Gygi SP, Rist B, Gerber SA, Turecek F, Gelb MH, Aebersold R (1999) Quantitative analysis of complex protein mixtures using isotope-coded affinity tags. Nat Biotechnol 17(10): 994–999
12. Bantscheff M, Schirle M, Sweetman G, Rick J, Kuster B (2007) Quantitative mass spectrometry in proteomics: a critical review. Anal Bioanal Chem 389(4):1017–1031. doi:10.1007/s00216-007-1486-6
13. Zhang G, Ueberheide BM, Waldemarson S, Myung S, Molloy K, Eriksson J, Chait BT, Neubert TA, Fenyö D (2010) Protein quantitation using mass spectrometry. Meth Mol Biol (Clifton, NJ) 673:211–222. doi:10.1007/978-1-60761-842-3_13
14. Keiji K, Takashi I (2008) Mass spectrometry-based approaches toward absolute quantitative proteomics. Curr Genomics 9(4):263–274. doi:10.2174/138920208784533647
15. Domon B, Aebersold R (2010) Options and considerations when selecting a quantitative proteomics strategy. Nat Biotechnol 28(7): 710–721
16. Iliuk A, Galan J, Tao WA (2009) Playing tag with quantitative proteomics. Anal Bioanal Chem 393(2):503–513. doi:10.1007/s00216-008-2386-0
17. Geiger T, Wisniewski JR, Cox J, Zanivan S, Kruger M, Ishihama Y, Mann M (2011) Use of

stable isotope labeling by amino acids in cell culture as a spike-in standard in quantitative proteomics. Nat Protoc 6(2):147–157
18. Oda Y, Huang K, Cross FR, Cowburn D, Chait BT (1999) Accurate quantitation of protein expression and site-specific phosphorylation. Proc Natl Acad Sci 96(12):6591–6596. doi:10.1073/pnas.96.12.6591
19. Ong S-E, Blagoev B, Kratchmarova I, Kristensen DB, Steen H, Pandey A, Mann M (2002) Stable Isotope labeling by amino acids in cell culture, SILAC, as a simple and accurate approach to expression proteomics. Mol Cell Proteomics 1(5):376–386. doi:10.1074/mcp.M200025-MCP200
20. Mann M (2006) Functional and quantitative proteomics using SILAC. Nat Rev Mol Cell Biol 7(12):952–958
21. Harsha HC, Molina H, Pandey A (2008) Quantitative proteomics using stable isotope labeling with amino acids in cell culture. Nat Protoc 3(3):505–516
22. Zhang R, Sioma CS, Wang S, Regnier FE (2001) Fractionation of isotopically labeled peptides in quantitative proteomics. Anal Chem 73(21):5142–5149. doi:10.1021/ac010583a
23. Hsu J-L, Huang S-Y, Chow N-H, Chen S-H (2003) Stable-isotope dimethyl labeling for quantitative proteomics. Anal Chem 75(24):6843–6852. doi:10.1021/ac0348625
24. Kovanich D, Cappadona S, Raijmakers R, Mohammed S, Scholten A, Heck AJR (2012) Applications of stable isotope dimethyl labeling in quantitative proteomics. Anal Bioanal Chem404(4):991–1009.doi:10.1007/s00216-012-6070-z
25. Boersema PJ, Raijmakers R, Lemeer S, Mohammed S, Heck AJR (2009) Multiplex peptide stable isotope dimethyl labeling for quantitative proteomics. Nat Protoc 4(4): 484–494
26. Chahrour O, Cobice D, Malone J (2015) Stable isotope labeling methods in mass spectrometry-based quantitative proteomics. J Pharm Biomed Anal 113:2–20. doi:10.1016/j.jpba.2015.04.013
27. Karp NA, Huber W, Sadowski PG, Charles PD, Hester SV, Lilley KS (2010) Addressing accuracy and precision issues in iTRAQ quantitation. Mol Cell Proteomics 9(9):1885–1897
28. Reynolds KJ, Yao X, Fenselau C (2002) Proteolytic 18O labeling for comparative proteomics: evaluation of endoprotease Glu-C as the catalytic agent. J Proteome Res 1(1): 27–33. doi:10.1021/pr0100016
29. Heller M, Mattou H, Menzel C, Yao X (2003) Trypsin catalyzed 16O-to-18O exchange for comparative proteomics: tandem mass spectrometry comparison using MALDI-TOF, ESI-QTOF, and ESI-ion trap mass spectrometers. J Am Soc Mass Spectrom 14(7):704–718. doi:10.1016/S1044-0305(03)00207-1
30. Miyagi M, Rao KCS (2007) Proteolytic 18O-labeling strategies for quantitative proteomics. Mass Spectrom Rev 26(1):121–136. doi:10.1002/mas.20116
31. Kalra H, Adda CG, Liem M, Ang CS, Mechler A, Simpson RJ, Hulett MD, Mathivanan S (2013) Comparative proteomics evaluation of plasma exosome isolation techniques and assessment of the stability of exosomes in normal human blood plasma. Proteomics 13(22): 3354–3364. doi:10.1002/pmic.201300282
32. Abdallah C, Dumas-Gaudot E, Renaut J, Sergeant K (2012) Gel-based and gel-free quantitative proteomics approaches at a glance. Int J Plant Genomics 2012:17. doi:10.1155/2012/494572
33. Yan W, Chen SS (2005) Mass spectrometry-based quantitative proteomic profiling. Brief Funct Genomic Proteomic 4(1):27–38. doi:10.1093/bfgp/4.1.27
34. Wang M, You J, Bemis KG, Tegeler TJ, Brown DPG (2008) Label-free mass spectrometry-based protein quantification technologies in proteomic analysis. Brief Funct Genomic Proteomic 7(5):329–339. doi:10.1093/bfgp/eln031
35. Megger DA, Bracht T, Meyer HE, Sitek B (2013) Label-free quantification in clinical proteomics. Biochim Biophys Acta 1834(8):1581–1590. doi:10.1016/j.bbapap.2013.04.001
36. Wasinger VC, Zeng M, Yau Y (2013) Current status and advances in quantitative proteomic mass spectrometry. Int J Proteomics 2013:12. doi:10.1155/2013/180605
37. Neilson KA, Ali NA, Muralidharan S, Mirzaei M, Mariani M, Assadourian G, Lee A, van Sluyter SC, Haynes PA (2011) Less label, more free: approaches in label-free quantitative mass spectrometry. Proteomics 11(4):535–553. doi:10.1002/pmic.201000553
38. Arike L, Peil L (2014) Spectral counting label-free proteomics. In: Martins-de-Souza D (ed) Shotgun proteomics: methods and protocols. Springer, New York, NY, pp 213–222. doi:10.1007/978-1-4939-0685-7_14
39. Shinoda K, Tomita M, Ishihama Y (2010) emPAI Calc--for the estimation of protein abundance from large-scale identification data by liquid chromatography-tandem mass

spectrometry. Bioinformatics 26(4):576–577. doi:10.1093/bioinformatics/btp700

40. Chiu C-W, Chang C-L, Chen S-F (2012) Evaluation of peptide fractionation strategies used in proteome analysis. J Sep Sci 35(23): 3293–3301. doi:10.1002/jssc.201200631
41. Mathivanan S, Ji H, Tauro BJ, Chen YS, Simpson RJ (2012) Identifying mutated proteins secreted by colon cancer cell lines using mass spectrometry. J Proteomics 76:141–149. doi:10.1016/j.jprot.2012.06.031
42. Gangoda L, Keerthikumar S, Fonseka P, Edgington LE, Ang CS, Ozcitti C, Bogyo M, Parker BS, Mathivanan S (2015) Inhibition of cathepsin proteases attenuates migration and sensitizes aggressive N-Myc amplified human neuroblastoma cells to doxorubicin. Oncotarget 6(13):11175–11190. doi:10.18632/oncotarget.3579
43. Keerthikumar S, Gangoda L, Liem M, Fonseka P, Atukorala I, Ozcitti C, Mechler A, Adda CG, Ang CS, Mathivanan S (2015) Proteogenomic analysis reveals exosomes are more oncogenic than ectosomes. Oncotarget 6: 15375–15396
44. Paoletti AC, Parmely TJ, Tomomori-Sato C, Sato S, Zhu D, Conaway RC, Conaway JW, Florens L, Washburn MP (2006) Quantitative proteomic analysis of distinct mammalian mediator complexes using normalized spectral abundance factors. Proc Natl Acad Sci U S A 103(50):18928–18933. doi:10.1073/pnas.0606379103
45. McIlwain S, Mathews M, Bereman MS, Rubel EW, MacCoss MJ, Noble WS (2012) Estimating relative abundances of proteins from shotgun proteomics data. BMC Bioinformatics 13:308. doi:10.1186/1471-2105-13-308

Chapter 5

TMT One-Stop Shop: From Reliable Sample Preparation to Computational Analysis Platform

Mehdi Mirzaei, Dana Pascovici, Jemma X. Wu, Joel Chick, Yunqi Wu, Brett Cooke, Paul Haynes, and Mark P. Molloy

Abstract

In this chapter we describe the workflow we use for labeled quantitative proteomics analysis using tandem mass tags (TMT) starting with the sample preparation and ending with the multivariate analysis of the resulting data. We detail the step-by-step process from sample processing, labeling, fractionation, and data processing using Proteome Discoverer through to data analysis and interpretation in the context of a multi-run experiment. The final analysis and data interpretation rely on an R package we call *TMTPrepPro*, which are deployed on a local GenePattern server, and used for generating various outputs which are also outlined herein.

Key words Quantitative shotgun proteomics, TMT, Software workflow

1 Introduction

Enabled by developments in mass spectrometers with electrospray ionization [1] sources and advancements in liquid chromatography (LC) systems, as well as completion of the genome sequences in a number of organisms, shotgun proteomics using tandem MS found its popularity and has had a substantial impact in many proteomics fields. LC based quantitative proteomics are classified into two categories: label-free (spectral counting, area under the curve), and stable isotope labeling (chemical and metabolic) [2–4]. Label free approaches, as the name suggests do not use any label and typically have a wide dynamic range and high analytical depth. In contrast, stable isotope labeling techniques incorporate differential isotope mass tags into the proteomes of experimental samples. While metabolic labeling requires actively dividing cells to incorporate the tag, chemical derivatization offers the advantage of labeling free amines in proteins regardless of metabolic state. The chemical tags affect only the mass of the labeled peptides while not

Shivakumar Keerthikumar and Suresh Mathivanan (eds.), *Proteome Bioinformatics*, Methods in Molecular Biology, vol. 1549, DOI 10.1007/978-1-4939-6740-7_5, © Springer Science+Business Media LLC 2017

altering their relative physiochemical properties. As the stable isotope tags of differing mass can be mixed, a significant advantage compared with label-free quantitation is that the differentially labeled peptides can be co-fractionated and co-separated during LC analysis while being simultaneously detected by mass spectrometry [5].

In stable isotope labeled approaches, labeled peptides are distinguished from each other via their mass shift signatures; hence, relative and absolute quantitation are achieved by comparison of the ion chromatogram peak areas of heavy and light labeled peptides, usually measured simultaneously in the same biological sample. Various chemical derivatization approaches using stable isotope labeling have been reported including ICAT [6], dimethyl labeling [7], iTRAQ [5] and TMT [8], while the most common metabolic labeling methods are ^{15}N labeling [9] and SILAC [10].

The major advantage of the iTRAQ and TMT techniques lies in their multiplexing capabilities; the relative protein abundance of many different samples can be determined at same time, which makes them an attractive option for larger scale experiments that would otherwise be prohibitive due to cost and/or instrument time. For example, iTRAQ is able to analyze up to eight samples (iTRAQ-8-plex), while TMT allows analysis of up to ten (TMT-10plex) samples simultaneously in a single MS run. However, alternative strategies for the synthesis of these multiplexing reagents have demonstrated the feasibility of analyzing more than ten samples in a single MS run [11]. These reagents share the same reactive group, N-hydroxy-succinimide (NHS), which has a high specificity for primary amine groups (α- and ϵ-amino groups) of peptides. The reaction is rapid and highly specific, and it irreversibly labels the free amines located at the N-terminus and the amine groups of the lysine side chain in proteins and peptides. Hence, labeling is not limited to a subset of proteins and all tryptic peptides will contain at least one site of modification. The tagging reagents are designed so that labeled peptides from the different starting materials are isobaric and co-elute in HPLC. Therefore, unlike metabolic labeling approaches, quantitation is not performed at the MS1 level, but rather using fragment ion tandem mass spectra. For labeled peptides, the mass tags are released during peptide fragmentation to produce "reporter" fragment ions—one for each biological sample. The intensity of these reporter ions corresponds to the relative peptide abundances present in the different samples, which are then summarized at the protein level based on peptide to protein sequence assignments. Since all tryptic peptides are isotopically labeled, more than one peptide will represent a protein which could potentially be identified, and thus the technique is likely to provide more accurate identification and quantification than the label free approach [3].

Apart from their superior multiplexing capability, isobaric labeling approaches present unique challenges, such as the cost of reagents and also a requirement for high-resolution mass spectrometer to analyze the diagnostic fragment ions. Moreover the accuracy and precision of the quantitative data can be compromised due to co-elution of multiple peptides within the isolation window that is selected for MS2 fragmentation. These interferences cause bias in the reporter ion intensities, often leading to reporter ion ratio compression and narrowing of the linear quantitative range [12]. Several approaches have been suggested to mitigate this issue. One practical solution is to reduce sample complexity by pre-fractionating (e.g., by strong cation exchange chromatography) the labeled and pooled peptides prior to LC-MS. Savitski et al. reported that using delayed peptide fragmentation closer to the apex of the chromatographic peak in LC-MS/MS results in twofold reduction in co-fragmentation, and hence a significant improvement in quantification. Finally, Ting et al. [13] demonstrated that the ratio compression/interference effect can be largely mitigated by performing an additional isolation and fragmentation event (MS3 scan). These methods were further improved to incorporate multiple notched waveforms that boost the available signal generated by the reporter ions, which subsequently leads to more reliable quantitation [14]. Wenger et al also demonstrated the use of gas-phase purification to improve precursor ion isolation selectivity [15].

Regardless of the quantitative method used, requirements for a successful quantitative proteomic analysis are the use of sufficient biological replicates, efficient sample preparation, appropriate MS instruments, and, finally, reliable bioinformatics tools and workflows for statistical data analysis. In this chapter we describe the steps we undertake in quantitative proteomics experiments using TMT, from a sample preparation we have found efficient and reliable over the course of numerous experiments, and through to the analysis pipeline that enables a quick first look at a data in a multivariate, experimentally relevant way. The details include critical steps required to achieve high labeling efficiency, export of data from Proteome Discoverer for analysis using our *TMTPrepPro* software package, data interpretation, and subsequent steps that can be undertaken to further place the data in the context of pathways and biological processes.

2 Materials

2.1 Sample Preparation

1. pH meter.
2. pH paper.
3. Reducing agent (5 mM DTT in Milli-Q water).

4. Alkylating agent (10 mM iodoacetamide in Milli-Q water).
5. Methanol (LC-MS grade).
6. Chloroform.
7. Acetone.
8. Acetonitrile (LC-MS grade).
9. 8 M urea in 50 mM Tris (pH 8.8).
10. BCA assay (Thermo Scientific, Rockford, IL).
11. Lys-C (Wako, Japan).
12. Trypsin (Promega, Madison, WI).
13. Incubator.
14. 130 mg solid-phase extraction cartridge (Sep-Pak, Waters, Milford, MA).
15. Trifluoroacetic acid (Sigma-Aldrich).
16. Formic acid (Sigma-Aldrich).
17. 4-(2-Hydroxyethyl)piperazine-1-ethanesulfonic acid—HEPES (Sigma-Aldrich).
18. MicroBCA assay (Thermo Scientific, Rockford, IL).
19. 10plex TMT reaction (Thermo, San Jose, CA).
20. Anhydrous acetonitrile (Thermo Scientific).
21. 5% hydroxylamine (Sigma-Aldrich).
22. Empore SDB-RPS disks (3M-Empore).
23. 16 G needle (Hamilton).
24. Syringe plunger 100 μl (Hamilton).
25. 5% ammonium hydroxide in 80% acetonitrile.
26. Q Exactive Orbitrap mass spectrometer (Thermo Scientific).
27. Proteome Discoverer V1.3 (Thermo Scientific).
28. *TMTPrepPro* (In-house R package which can be downloaded from ftp://ftp.proteome.org.au/TMTPrepPro).

3 Methods

In our laboratory we employ a number of quantitative proteomics approaches such as label free shotgun proteomics, iTRAQ, TMT, and SWATH [16, 17]. We have found isobaric labeling is applicable across a wide range of biological samples, including those derived from animals, humans, and plants, due to fact the labels are incorporated after the extraction and digestion steps. Nonetheless, to establish a reliable isobaric tagging workflow, several aspects of the widely used label free workflows require modification, particularly during sample preparation and data acquisition steps.

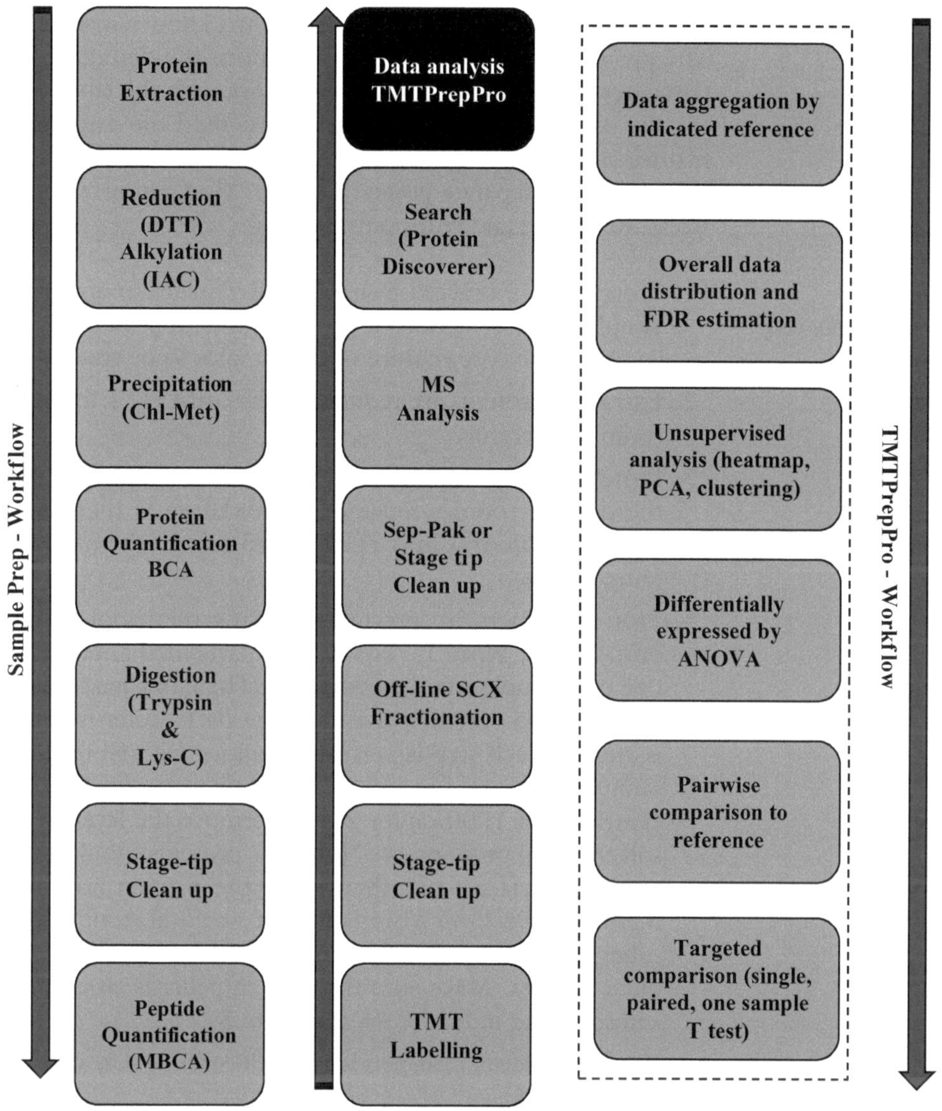

Fig. 1 TMT workflow

Extraneous primary amine groups, for example in the commonly used ammonium bicarbonate buffer, would hinder the labeling efficiency. Overall, consideration must be given to primary amine free reagents, digestion with multiple enzymes for enhancing sequence coverage, and protein and peptide cleaning steps before and after labeling step.

The workflow described below (Fig. 1) for sample preparation and data analysis can be applied to any isobaric labeling experiment. The only step which might be different for each study is the protein extraction step. Regardless of the extraction procedure used, samples will need to be reduced and alkylated prior to precipitation and the rest of steps in the workflow remain the same. The protein precipitation step will remove the detergent or contaminates, which might cause interference with the labeling.

For the figures in this chapter we used an experiment on mature rice leaves exposed to drought stress over a time course containing five separate points: Control, Moderate stress, Extreme stress, Recovery, and end-point Control.

3.1 Protein Extraction, Reduction, Alkylation, Precipitation, Quantification, and Digestion

1. Proteins are extracted from the cells/tissues or any biological samples. The selection of protein extraction method is solely based on the type/nature of the samples to be studied.
2. Extracted proteins are reduced with 5 mM DTT for 15 min at room temperature.
3. Protein samples are then alkylated with 10 mM for 30 min in the dark at room temperature. Alkylation is then quenched with addition of 5 mM DTT for 15 min in the dark at room temperature.
4. Protein extracts are precipitated using methanol–chloroform protocol (*see* **Note 1**). Firstly, four parts methanol is added to one part sample and then vortexed. Then, one part chloroform is added and vortexed. Lastly, three parts water is added and vortexed. Each step is performed on ice with all reagents and samples are stored in ice. Let the mix stand for 5 mins then centrifuge at $1000 \times g$ for 2 min. Remove the layer of organic solvent from proteins (*see* **Note 2**)—proteins should aggregate at the interface—and then wash the protein with ice-cold methanol. Perform a wash with ice-cold acetone. Leave the tubes with lids open in fume hood for 10 min, to achieve complete dryness. Make sure the protein pellet is air-dried before resuspending in 8 M Urea (*see* **Note 3**).
5. Protein pellet is resuspended in 200 μl 8 M Urea in 50 mM Tris (pH 8.8).
6. Protein concentration is determined by BCA assay using bovine serum albumin (BSA) as a standard.
7. Samples are then digested with Lys-C (Wako, Japan) at a 1:100 enzyme–protein ratio overnight at room temperature.
8. Samples are then diluted with 50 mM Tris pH 8.8 to a final concentration <2 M urea.
9. Digested proteins are further digested with Trypsin (Promega, Madison, WI) at a 1:100 enzyme–protein ratio for at least 4 h at 37 °C (*see* **Note 4**). Samples are then acidified with TFA to a final concentration of 1 % (check with pH strip: pH 2–3).
10. Samples are then desalted using SDB-RPS (3M-Empore) Stage Tips.

3.2 SDB-RPS Desalting Using Stage Tips

1. Each sample would require a separate Stage Tips (*see* **Note 5**). To make a Stage Tip, first cut and stack layer/s of SDB-RPS disks using a 16-G Hamilton syringe needle into bottom of 200 μl tip. Each disk binds 20 μg of peptide, in case more capacity is required increase number of disks or alternatively use a larger diameter, 14-G needle so that each disk has the capacity to bind 30 μg peptide (*see* **Note 6**).
2. Acidify the peptide samples using TFA to pH 2–3 (*see* **Note** 7).
3. Place Stage Tips into collection tubes, load the samples on top of Stage Tips and centrifuge at 1000–2000 × *g* until all solution passed through the Stage Tips (*see* **Note 8**). Optional: flow-through can be collected into separate tubes and stored.
4. Wash the Stage Tip twice with 100 μl of wash buffer (0.2% TFA) and centrifuge at 1000 × *g*, empty the collecting tubes if needed.
5. Elute the peptides with 100 μl of elution buffer (5% ammonium hydroxide/80% ACN), dry the eluent using a vacuum centrifuge.

3.3 TMT Labeling

1. Dried peptides are resuspended in 200 μl of 200 mM HEPES—pH 8.8 (*see* **Note 9**).
2. Peptide concentration is measured using MicroBCA (Thermo Scientific, Rockford, IL), 70 μg from each samples are aliquoted for labeling in a 10plex TMT reaction (Thermo, San Jose, CA).
3. Add 41 μl of anhydrous acetonitrile to each 0.8 mg label vial, followed by occasional vortexing for 5 min and brief centrifugation (*see* **Note 10**).
4. Ten TMT labels (ten labels) are added to the ten individual protein samples. Labeling is performed at room temperature for 1 h with occasional vortexing.
5. Add 8 μl of 5% hydroxylamine to each sample, vortex and incubate at RT for 15 min (Note). (Quenching removes TMT label from tyrosines.)
6. All ten labeled samples are combined in a clean 2 ml Eppendorf tube. The combined mixture of TMT labeled peptides is dried down using speed vacuum centrifuge.
7. Dried peptide mixture is reconstituted in 1% TFA (pH around 2–3). The mixture is desalted using on a 130 mg solid-phase extraction (Sep-Pak, Waters, Milford, MA) and then again dried down using speed vacuum centrifuge.

3.4 C18 Desalting Using Sep-Pak

1. Attach a 3 ml luer-lok syringe (without the plunger) to a Sep-Pak cartridge. Place the syringe-cartridge assembly vertically on top of the 15 ml falcon tube for sample collection.

2. Wash the Sep-Pak with 2 ml of 100% methanol by pipetting Methanol into the syringe, attach the plunger, and apply pressure slowly to pass the methanol completely through.
3. Wash the Sep-Pak with 2 ml 80% acetonitrile, 0.5% acetic acid in Milli-Q water as above.
4. Load the sample, add another 2 ml of 1% FA in Milli-Q water, apply gentle pressure to let sample pass through at a slow rate (approx. 0.5 ml/min). Optional: flowthrough can be collected in a separate tube for further inspection of unbound peptides.
5. Wash the Sep-Pak with 2 ml of 1% FA in Milli-Q water as above.
6. Elute peptides into 2 ml Eppendorf tube with 1.8 ml of 80% acetonitrile, 0.5% acetic acid in Milli-Q water. Dry the eluents using speed vacuum centrifuge.

3.5 Offline SCX Fractionation

1. Offline SCX fractionation is carried out, to reduce the complexity of the mixture, using an Agilent 1260 quaternary HPLC pump with a PolyLC polysulfoethyl aspartamide column (200 mm×2.1 mm, 5 μm, 200 Å; PolyLC, Columbia, MD) and UV detection at 210 nm.
2. The column is equilibrated with buffer A (5 mM KH_2PO_4, 25% v/v acetonitrile (ACN), pH 2.72), which is also used for sample resuspension, sample injection and peptide adsorption to the column. Peptide elution is achieved with a linear gradient of 10–45% buffer B (5 mM KH2PO4, pH 2.72, 350 mM KCl, 25% ACN) for 70 min, which is then rapidly increased from 45 to 100% buffer B for 10 min at a flow rate of 300 μl/min.
3. A total of 36 fractions of varying volumes are collected in a 96-well collection plate and dried down by vacuum centrifugation. 100 μl 1% TFA is added to each of 36 wells (wells containing peptides) and vortexed well for 10 min at 4 °C, before being combined into 12 fractions based on UV absorbance.
4. These 12 fractions are desalted using SDB-RPS Stage Tips, dried down using a vacuum centrifuge and reconstituted in 0.1% formic acid in preparation for LC-MS/MS.

3.6 Nanoflow LC-MS/MS for TMT Labeling Samples

1. Samples are analyzed on a Q Exactive Orbitrap mass spectrometer (Thermo Scientific) coupled to an EASY-nLC1000 (Thermo Scientific).
2. Reversed-phase chromatographic separation is carried out on a 75 μm id.×100 mm, C18 HALO column, 2.7 μm bead size, 160 Å pore size.
3. A linear gradient of 1–30% solvent B (99.9% ACN/0.1% FA) is run over 170 min. The mass spectrometer is operated in the data-dependent mode to automatically switch between Orbitrap MS and ion trap MS/MS acquisition.

4. Survey full scan MS spectra (from m/z 350–1850) are acquired at precursor isolation width of 0.7 m/z, resolution of 70,000 at m/z 400 and an AGC (Automatic Gain Control) target value of 1 × 106 ions.
5. For identification of TMT labeled peptides, the ten most abundant ions are selected for higher energy collisional dissociation (HCD) fragmentation. HCD normalized collision energy is set to 35 % and fragmentation ions are detected in the Orbitrap at a resolution of 70,000.
6. Target ions that have been selected for MS/MS are dynamically excluded for 90 s. For accurate mass measurement, the lock mass option is enabled using the polydimethylcyclosiloxane ion (m/z 445.12003) as an internal calibrant.

3.7 Data Processing

The relative abundance of peptides is determined by calculating the ratios of the of the reporter ions intensities; subsequently, the integration of relative quantification at the peptides level would represent the relative expression of the proteins. Software such as Proteome Discoverer (Thermo Scientific), MaxQuant [18], or Mascot can be used to generate the protein fold changes required for analysis. We use Proteome Discoverer as our main data analysis platform. Once the quantitative protein ratios are generated, we use our own R software workflow to implement the subsequent multivariate analysis steps commonly needed for more complex experiments. Our simple strategy for analyzing labeled experiments that include multiple runs relies on using a common reference to combine ratios from disparate runs, which in the case of the TMT 10-plex is not too stringent a requirement. We must emphasize that many other approaches are possible, such as more sophisticated statistical approaches [19, 20] or the ProteoIQ software suite [21]. While the multivariate analysis steps we automate are intended to cover some commonly used scenarios, there will be some experiments that have to be analyzed in a customized fashion, for instance those having sample pairing (such as treated/control of the same patient or cell line). Hence, we designed a targeted approach where the user has the option to choose the specific ratios and statistical tests carried out.

1. Raw data files generated by Xcalibur software (Thermo Scientific) are processed using Proteome Discoverer V1.3 (Thermo Scientific) and a local MASCOT server (version 2.3; Matrix Science, London, UK).
2. The MS/MS spectra are searched against the protein NCBI Rice database. The MS tolerance is set to ±10 ppm and the MS/MS tolerance to 0.1 Da and Trypsin with one missed cleavage.

3. Carbamidomethylation of cysteine and 10-plex TMT tags on lysine residues and peptide N-termini are set as a static modification, while oxidation of methionine and deamidation of asparagine and glutamine residues are set as a variable modification.
4. Search result filters are selected as follows: only peptides with a score >15 and below the Mascot significance threshold filter of $p = 0.05$ are included and single peptide identifications required a score equal to or above the Mascot identity threshold.
5. Protein grouping is enabled such that when a set of peptides in one protein are equal to, or completely contained, within the set of peptides of another protein, the two proteins are contained together in a protein group.
6. Proteins with at least two unique peptides are regarded as confident identifications. Relative quantitation of proteins is achieved by pairwise comparison of TMT reporter ion intensities, for example, the ratio of the labels for each of treatment replicates (numerator) versus the labels of their corresponding control replicates (denominator).

3.8 TMT Data Analysis Program TMTPrepPro: Uploading the Data

The *TMTPrepPro* scripts are implemented in the R programming language and are available as an R package, which is accessed in our group through a graphical user interface provided via a local GenePattern [1] server. There are two distinct analyses types: overall multivariate analysis, and targeted pairwise comparisons. The overall multivariate analysis combines ratios from a number of runs with respect to an indicated reference, performs unsupervised analyses such as clustering and PCA, determines differentially expressed proteins by an ANOVA approach, and carries out pairwise comparisons detected automatically based on the experimental design. The targeted pairwise comparison is defined for a single run only at this point, and can be used to enter specific comparisons of interest and the analysis suitable for them.

As with all labeled experiments, a key input is the experimental design showing the placing of samples on runs, which needs to be created first as an Excel spreadsheet; an example is given in Table 1.

The user interface requires the Proteome Discoverer data to be uploaded and a few parameters to be set as follows:

1. Upload the protein search results extracted as tab-delimited files at the previous step; if multiple runs are used the files should be zipped up together.
2. Upload the design file describing the experimental group for each label in the first tab, and the label to be used as reference in the second tab. *See* Table 1 for an example (*see* **Note 11**).
3. Set the limits to be used for cutoffs for differential expression (by default 1.5), for number of counts per peptide (by default

Table 1
Design Excel spreadsheet example for overall TMT job

Tab1				
Label	*Replicate*	*Group*	*Replicate*	*Group*
126	C1	1Control	C1	1Control
127_N	C2	1Control	C2	1Control
127_C	M1	2Moderate	M1	2Moderate
128_N	M2	2Moderate	M2	2Moderate
128_C	E1	3Extreme	E1	3Extreme
129_N	E2	3Extreme	E2	3Extreme
129_C	R1	4Recovery	R1	4Recovery
130_N	R2	4Recovery	R2	4Recovery
130_C	C1After	5ControlA	C1After	5ControlA
131	C2After	5ControlA	C2After	5ControlA
Tab2				
File				*Use reference*
Run1 ProtDisc Output File.txt				126
Run2 ProtDisc Output File.txt				126

no limit), and for the protein ratio z-score, defined as $100 \times \log(\text{ratio})/\text{Variability}$ (by default 2).

4. Start the analysis; several calculations will be performed and various spreadsheets and images will be generated as described in detail below.

3.9 TMTPrepPro Outputs for the Overall Multivariate Analysis Job

The TMT overall multi-run job yields a multivariate overview of the data and can be divided into several analyses categories: data aggregation and summaries, overall data quality and FDR based on replicates (if they exist), unsupervised analyses (clustering and PCA), ANOVA, and pairwise comparisons to the common reference.

3.9.1 Data Aggregation with Respect to Indicated References

(a) *ResultsOverall.xlsx*
The spreadsheet contains the combined ratios, variabilities, and counts, alongside results from other statistical analyses which are described later.

3.9.2 Overall Data Distribution and FDR Based on Replicates

(a) *BoxplotDensity.png*
An image is show in Fig. 2, showing the boxplots and density plots of all the log ratios extracted to the indicated ratio. In the shown image, the indicated reference is a control sample.

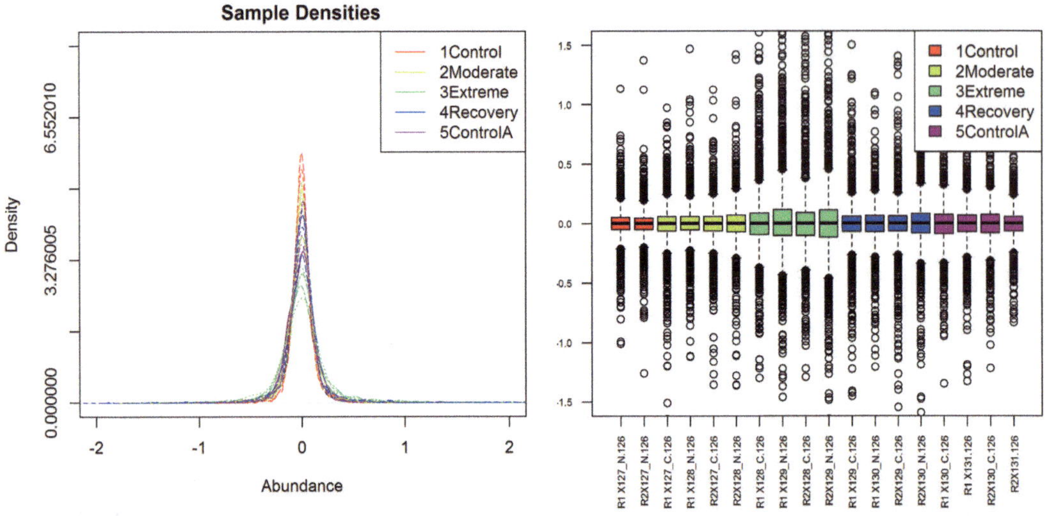

Fig. 2 Boxplots and density plots

The coloring of samples is done based on the Group of the top reference in the ratio, and the precise labels for the ratios are indicated in the boxplot. To interpret, one looks for similar patterns across the groups, and no sample standing out as having an unusual distribution.

(b) *Correlation heatmap.png*

The correlation matrix of all log ratios to the indicated reference is generated, and clustered as a heatmap. Ideally one looks for groups appearing close together, although this may not be the case if the differences between samples are small. In the case of multiple runs one looks for no clear clustering of the different runs together, which would indicate run effects that have to be accounted for.

(c) *FDR based on reference replicates images*

If reference replication is present in the design, then estimates of false discovery rates based on replicates will be generated as follows. Ratios with the same group as the reference will be identified, in this example other Control1 ratios (127 N has the same group as reference 126). For each such ratio, the number of proteins found differentially expressed based on three criteria will be determined: (1) ratio > ratio cutoff parameter (1.5 default); (2) ratio > ratio cutoff and peptide counts > 1; and (3) ratio > ratio cutoff and *z*-score > *z*-score cutoff (two default). The percentages of proteins identified as changing are listed in the image subtitle, and the plot of log ratios and counts and log ratios and log(absolute value of *z*-score) are plotted side by side. For a technical replicate or a close biological replicate, such as pools of plants in similar conditions, one would expect low FDR percentages around 1–2 %,

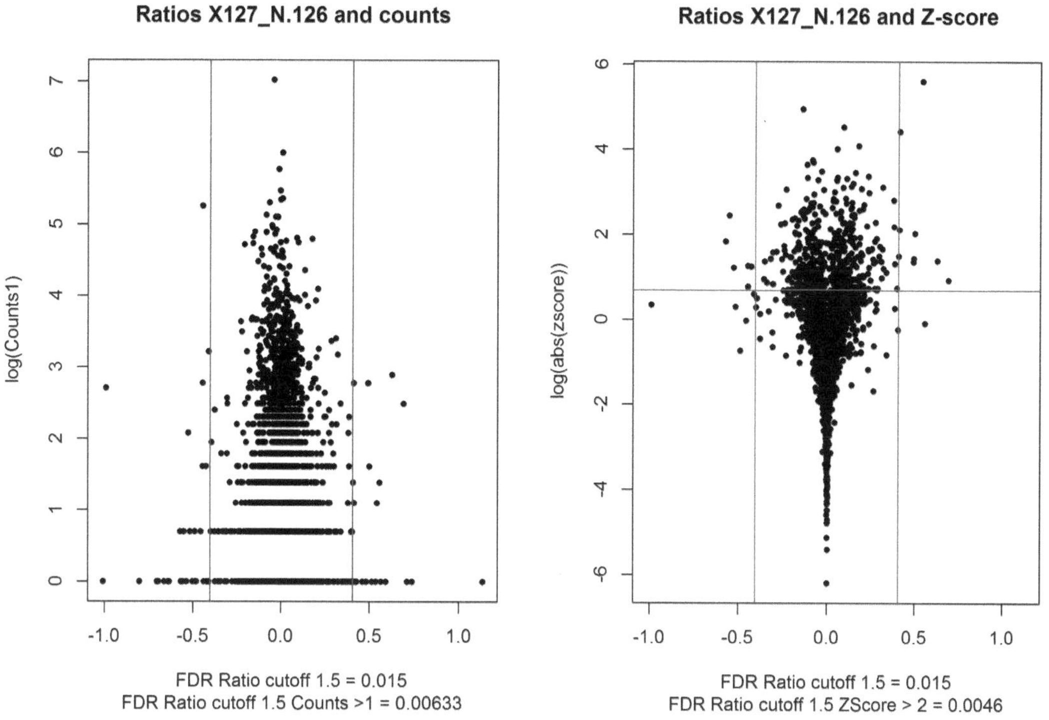

Fig. 3 FDR based on reference replicates

and even lower when counts or z-scores are taken into account. For a biological replicate, the percentage of proteins such identified could be a lot higher. The figures will be generated containing the number of the file and of the reference, for instance *File1Ref126Ratio1.png*; an example is shown in Fig. 3.

(d) *Ratio correlations by group images*

For each of the group categories present in the design, side-by-side plots of the ratios from the same category will be generated and maximum and minimum correlations will be included in the title. The figure file names will be automatically created to include the group category, for instance *Cor3Extreme.png*.

3.9.3 Unsupervised Analyses

(a) *HeatmapAll.png*

The log-transformed ratios extracted will be visualized on a heatmap, using the R implementation with complete linkage and correlation based distance; proteins with missing values are removed prior to clustering. The columns will be colored based on the group of the top reference.

(b) *PCA3dPlot.png*, *PCATopLoadings.png* and *PCATopLoadingsProteinPatterns.png*

The log-transformed ratios with missing values removed are also visualized using a principal component analysis (PCA). The three-dimensional plot of the PCA scores for the first three components is included in the *PCA3dPlot.png* image. The loadings image shows the top five proteins with the highest loadings in each of the three components. Finally, the protein pattern figure shows the protein patterns of the proteins with highest loadings across the experimental groups identified in the design. For example, in Fig. 4, the top 3D image shows the principal component scores, the middle barplot shows the proteins with the top loadings for each component, and the bottom boxplot shows the pattern over the experimental conditions of the five proteins with top loadings for principal component 1.

(c) *ResultsOverall.xlsx*
The spreadsheet described earlier also contains the protein loadings and principal component scores generated by the PCA analysis in two separate tabs, called *PCAScores* and *PCALoadings*. It can be used for instance to identify more than the top five proteins with highest loadings generated automatically in the images described above.

3.9.4 ANOVA

(a) *ResultsOverall.xlsx*
This spreadsheet described earlier, which contains the combined ratios and variabilities, also contains statistics generated while running a one way ANOVA analysis for each protein in the set, including the *p*-value, the adjusted *p*-value using the Benjamini–Hochberg correction for multiple testing,the Geometric means of all ratios for each condition with respect to the selected reference.

(b) *Heatmap—Anova DE.png*
A heat map representation of all differentially expressed proteins identified at the previous step; similar to the heatmap described previously generated for all ratios. An example is shown in Fig. 5.

(c) *ClusterPatterns.png*
An example is shown in Fig. 6. This is another representation of the differentially expressed proteins, clustered first into four clusters using hierarchical clustering, then plotted showing the means across all experimental conditions for all proteins in each cluster. The ordering of the group categories is done alphabetically (*see* **Note 12**).

3.9.5 Pairwise Comparisons to the Reference

(a) *ResultsPairwise.xlsx*
This spreadsheet contains the results from a number of pairwise comparisons that are automatically carried out given the design and choice of reference. For each group category all the ratios extracted with respect to the given reference are log-transformed and compared to 0 via a one-sample *t*-test. Proteins with a *t*-test *p*-value <0.05 and average fold change > cutoff are regarded as differentially expressed. It is important

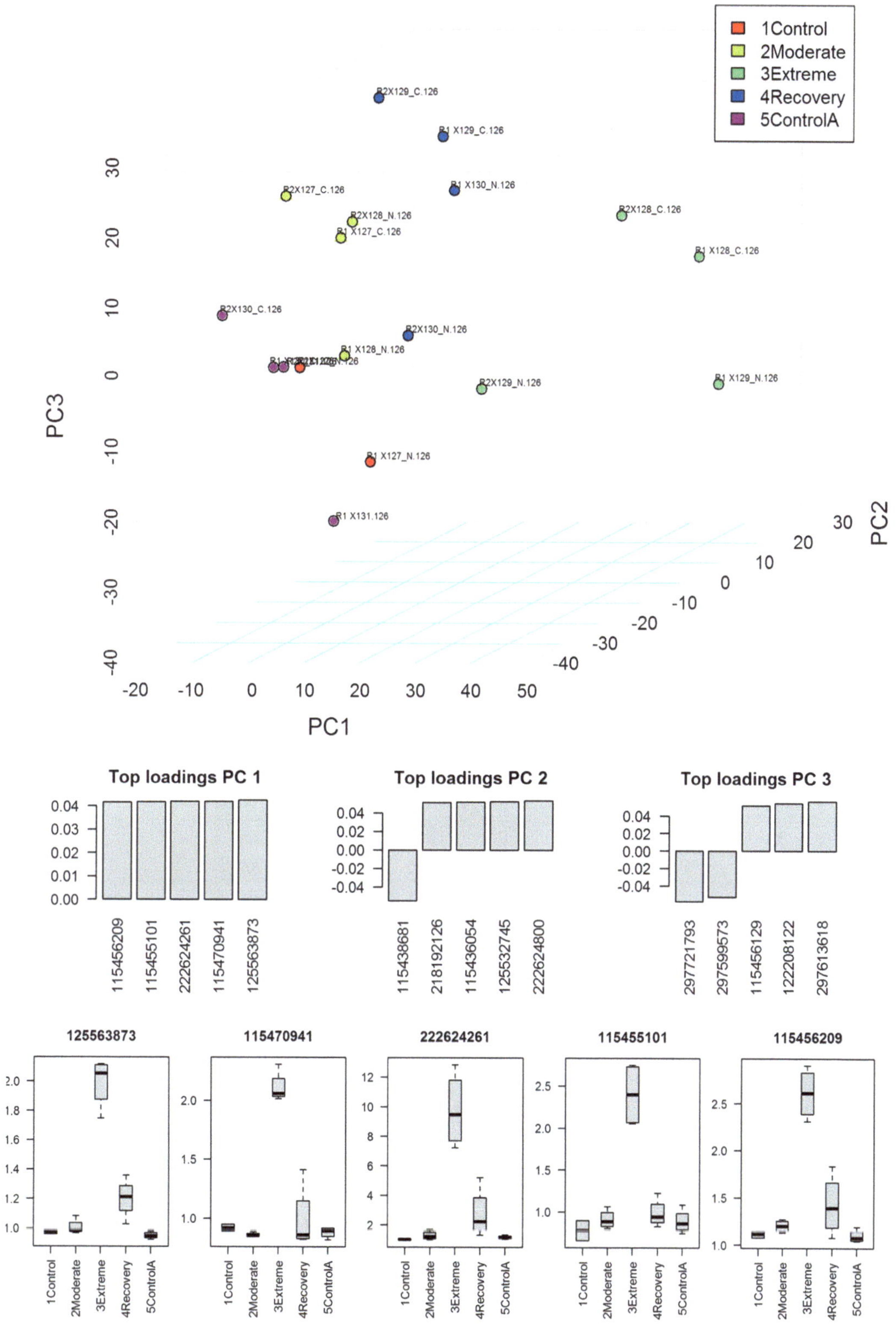

Fig. 4 PCA 3D plot, top loadings and protein patterns for the top loading proteins

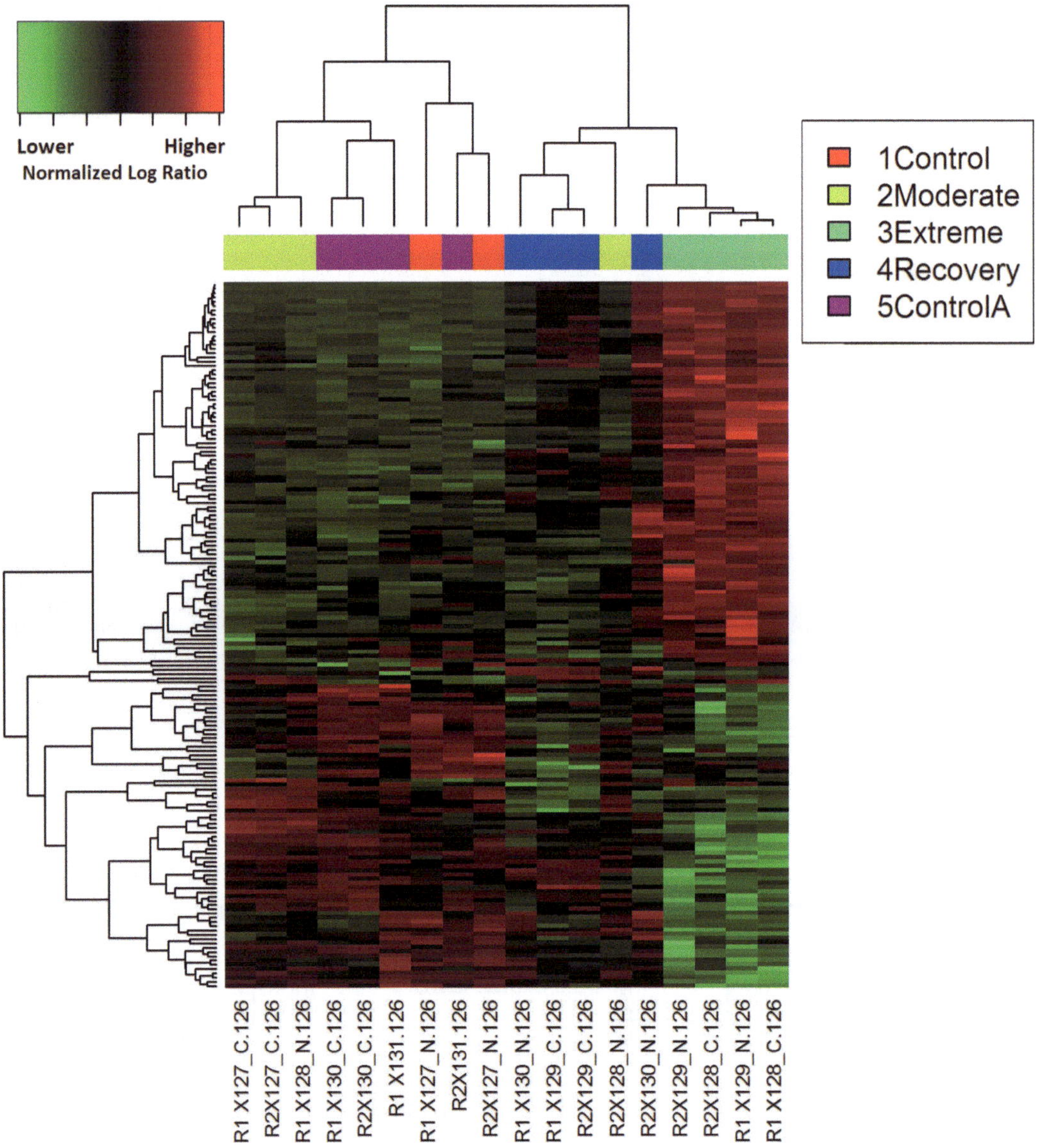

Fig. 5 Heatmap of differentially expressed proteins

to note that this set of pairwise comparisons is only relevant if the ratios are experimentally meaningful. For example, in the context of the five condition rice experiment, the references are control samples, and hence the pairwise comparisons extract differences between Control and control references, Moderate and control references, Extreme and control references, etc. If the reference is a common pool of all samples, then the comparisons are not easy to interpret and probably of no use.

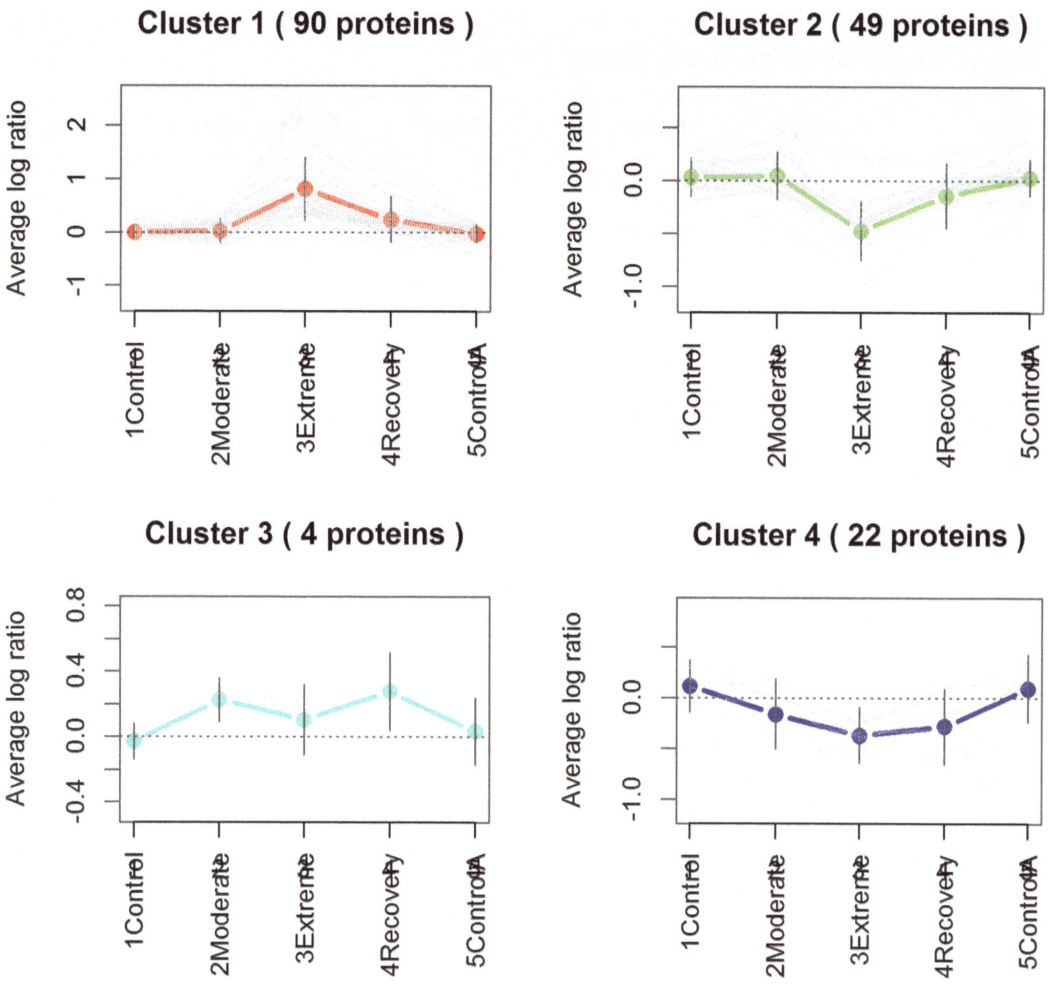

Fig. 6 Cluster patterns for differentially expressed proteins

(b) *Volcano plots images*
For each pairwise comparison undertaken above, the resulting *p*-values and log fold changes are visualized on volcano plots, one generated for each comparison, and labeled by the group category name, for instance "*Volcano 1Control.png*" (*see* **Note 13**).

(c) *Correlations of differentially expressed proteins by group images*
Side-by-side plots of the differentially expressed proteins for each comparison are also generated, showing the correlation of the respective ratios.

(d) *VennOverlap.png*
A Venn diagram showing the overlap of the differentially expressed proteins identified in the previous step is also generated; if more than three group levels are present then the three most numerous categories are used.

Table 2
Design Excel spreadsheet example for targeted TMT job

Label	Replicate	Group	Single	OneSampleTTest	Paired
126	C1	1Control	B	B	B1
127_N	C2	1Control	T	B	B2
127_C	M1	2Moderate			
128_N	M2	2Moderate			
128_C	E1	3Extreme		T	T1
129_N	E2	3Extreme		T	T2
129_C	R1	4Recovery			
130_N	R2	4Recovery			
130_C	C1After	5ControlA			
131	C2After	5ControlA			

3.10 TMTPrepPro Outputs from Targeted Analysis

The inputs into the *TMTPrepPro* for targeted analysis (Table 2) are similar to those into the overall multivariate analysis job described previously. All targeted analysis needs to be clearly specified in the design file; each requested comparison analysis is specified as a single column. The analysis specification includes the comparison type which can be one of single, paired and one sample *t*-test, and the labels involved in the comparison and their position (top or bottom) in the ratio. The comparison type is indicated as the column header and only the labels that will be used in the comparison are specified in the cell corresponding to that label. All other labels can be left blank. The specifications for labels are slightly different for different comparison types. For single comparison, two TMT labels need to be indicated as "T" (19) and "B" (bottom) respectively. For one sample *T* test, at least two labels need to be indicated as "T" and another two as "B". For paired comparison, exact two labels are marked as "T1" and "T2" respectively and another two as "B1" and "B2" respectively. An example of the design file can be found in the *TMTPrepPro* package. The details of each analysis are described below.

The outputs of the target analysis include a combined Results.xlsx and one image plot for each comparison defined in the design.

(a) Results.xlsx

The workbook contains multiple spreadsheets. The first sheet is named "Comparison" and contains the metadata of all the valid comparisons that have been executed. The second sheet is named as "ErrorNote" and contains the TMT labels specified in the design but do not exist in the TMT protein discoverer data. The other sheets contain the result data for each comparison.

(b) Vocano.png

For Single and OneSampleTTest comparison undertaken, the resulting *z*-score or *p*-value and log fold changes are visualized on volcano plots, one generated for each comparison and named by the comparison name such as *SingleComp 1 Volcano.png*. All differentially expressed proteins are highlighted as a red dot and labeled with their identification accessions.

(c) Paired.png

For Paired comparison, a scatter plot is generated to visually display the relationship between the ratios of the set of common proteins identified as differentially expressed consistently by the pair of single comparisons.

3.10.1 Single Comparison

The single comparison is specified as "Single" type in the design together with one TMT label indicated as "T" (19) and one as "B" (bottom). The comparison excludes proteins with missing ratio or variability. It performs a comparison of the specified labels for each proteins to 1 based on the ratio and *z*-score. The *z*-score is defined as $100 \times \log(\text{ratio})/\log(\text{variability})$. The default set of cutoffs for ratio and *z*-score are 1.5 and 2, respectively. The proteins that satisfies ratio > ratio cutoff AND *z*-score > *z*-score cutoff are classified as differentially expressed.

(a) Results.xlsx

The spreadsheet contains the ratio, peptide count and variability, *z*-score, and class for each protein. The class represents the classed that the protein detected as, which can be one of three values: −1, 0, and 1, representing downregulated, not differentially expressed and upregulated. To help visualization, all differentially expressed protein are highlighted, with upregulated as yellow and downregulated as blue.

(b) Volcano.png

A volcano plot of log *p*-value VS log ratio is generated, with all differentially expressed proteins labeled with their accession numbers and colored as red.

3.10.2 Paired

The paired comparison performs two single comparisons with the specified labels, T1/B1 and T2/B2, independently and then compare the consistence of the differentially expressed proteins identified. Here, consistent differentially expressed proteins means the set of proteins which are identified as upregulated or downregulated by both single comparisons.

(a) Results.xlsx

Similar to the result spreadsheet of Single comparison, the spreadsheet in Results.xlsx for Paired comparison contains two sets of ratios, peptide counts and variability, *z*-score, and class for each protein for the two sets of compared labels. Similarly, all differentially expressed proteins are highlighted.

(b) Paired.png
A scatter plot, ending in "Paired.png", visualize the relationship between the two ratios for each consistent differentially expressed proteins.

3.10.3 One Sample T Test

The one sample *T* test analysis takes four or more combined ratio labels (two top and two bottoms) from the design and extract all rations from the protein discoverer file. All the ratios are log-transformed and compared to 0 via a one sample *T* test. Proteins with a *t*-test *p*-value less than 0.05 and mean fold change greater than cutoff (1.5 default) are regarded as differentially expressed.

(a) Results.xlsx
The result spreadsheet contains all the extracted ratios, the mean ratios, the *p*-values, and the classes for all proteins.

(b) Volcano.png
A volcano plot of log *p*-value VS log mean ratio is generated, with all differentially expressed proteins labeled with their accession numbers and colored in red.

4 Notes

1. For the protein samples more than 200 μl, TCA/acetone precipitation method is recommended. This is mainly because, 2 ml Eppendorf tube will not be sufficient to fit all the reagents required for methanol–chloroform precipitation. Hence, the protein loss would be greater in a larger tubes (eg. 14 ml falcon tubes).
2. Do not disturb the protein pellet, do not intend to collect any solution below the formed pellet.
3. Air-dry the samples (tubes with caps off), leave the Eppendorf tubes on the bench for 5–10 min, make sure the pellet is not over-dried. If it turns brown, it would be difficult to bring back everything into solution.
4. Make sure the samples are diluted to lower than 2 M Urea and pH is maintained around 8.8. The high concentration of Urea prevents trypsin activity.
5. Despite C18 Empore disks, SDB-RPS disks do not require any activation or equilibration.
6. Stack SDB dics based on the amount of protein/peptide which needs to be cleaned. For instance, four dics are required for 120 μg (14 G needle).
7. Use high concentration of TFA (50%) to acidify the samples, start with 2 μl, vortex and check the pH using pH strips. Keep on adding until the pH reaches about 2 or 3. Do not let the

total volume go higher than 150 µl if 200 µl tips are used to stack disks on.

8. Peptide binding and elution steps should be done at 1000 × *g*, while the wash step can go up to 2000 × *g*.
9. Adjust the HEPES buffer to 8.8 using NaOH. Check pH with pH strips.
10. Remove the reagents from freezer just before the labeling step. Allow the TMT label vials reach room temperature before opening the lids.
11. This format can also be used to combine ratios with different reference labels (denominators) on the same run if appropriate.
12. On the cluster figure the group levels will be plotted in alphabetical order; add numbers 1, 2, 3, ... to the group labels to ensure plotting in the order that is meaningful to the experiment.
13. Only meaningful if the ratios to the reference have an experimental interpretation, for instance in this case "change from control"; usually meaningless if the common reference is a pool of all samples.

References

1. Reich M, Liefeld T, Gould J, Lerner J, Tamayo P, Mesirov JP (2006) GenePattern 2.0. Nat Genet 38(5):500–501
2. Yates JR, Ruse CI, Nakorchevsky A (2009) Proteomics by mass spectrometry: approaches, advances, and applications. Annu Rev Biomed Eng 11:49–79
3. Neilson KA, Ali NA, Muralidharan S, Mirzaei M, Mariani M, Assadourian G, Lee A, Van Sluyter SC, Haynes PA (2011) Less label, more free: approaches in label-free quantitative mass spectrometry. Proteomics 11(4):535–553
4. Patel VJ, Thalassinos K, Slade SE, Connolly JB, Crombie A, Murrell JC, Scrivens JH (2009) A comparison of labeling and label-free mass spectrometry-based proteomics approaches. J Proteome Res 8(7):3752–3759
5. Ross PL, Huang YN, Marchese JN, Williamson B, Parker K, Hattan S, Khainovski N, Pillai S, Dey S, Daniels S (2004) Multiplexed protein quantitation in Saccharomyces cerevisiae using amine-reactive isobaric tagging reagents. Mol Cell Proteomics 3(12):1154–1169
6. Gygi SP, Rist B, Gerber SA, Turecek F, Gelb MH, Aebersold R (1999) Quantitative analysis of complex protein mixtures using isotope-coded affinity tags. Nat Biotechnol 17(10): 994–999
7. Boersema PJ, Raijmakers R, Lemeer S, Mohammed S, Heck AJR (2009) Multiplex peptide stable isotope dimethyl labeling for quantitative proteomics. Nat Protoc 4(4): 484–494
8. Thompson A, Schäfer J, Kuhn K, Kienle S, Schwarz J, Schmidt G, Neumann T, Hamon C (2003) Tandem mass tags: a novel quantification strategy for comparative analysis of complex protein mixtures by MS/MS. Anal Chem 75(8):1895–1904
9. Oda Y, Huang K, Cross FR, Cowburn D, Chait BT (1999) Accurate quantitation of protein expression and site-specific phosphorylation. Proc Natl Acad Sci 96(12):6591–6596
10. Ong S-E, Blagoev B, Kratchmarova I, Kristensen DB, Steen H, Pandey A, Mann M (2002) Stable isotope labeling by amino acids in cell culture, SILAC, as a simple and accurate approach to expression proteomics. Mol Cell Proteomics 1(5):376–386
11. Braun CR, Bird GH, Wühr M, Erickson BK, Rad R, Walensky LD, Gygi SP, Haas W (2015) Generation of multiple reporter ions from a single isobaric reagent increases multiplexing capacity for quantitative proteomics. Anal Chem 87(19):9855–9863. doi:10.1021/acs.analchem.5b02307

12. Savitski MM, Sweetman G, Askenazi M, Marto JA, Lang M, Zinn N, Bantscheff M (2011) Delayed fragmentation and optimized isolation width settings for improvement of protein identification and accuracy of isobaric mass tag quantification on Orbitrap-type mass spectrometers. Anal Chem 83(23):8959–8967
13. Ting L, Rad R, Gygi SP, Haas W (2011) MS3 eliminates ratio distortion in isobaric multiplexed quantitative proteomics. Nat Methods 8(11):937–940
14. McAlister GC, Nusinow DP, Jedrychowski MP, Wühr M, Huttlin EL, Erickson BK, Rad R, Haas W, Gygi SP (2014) MultiNotch MS3 enables accurate, sensitive, and multiplexed detection of differential expression across cancer cell line proteomes. Anal Chem 86(14): 7150–7158. doi:10.1021/ac502040v
15. Wenger CD, Lee MV, Hebert AS, McAlister GC, Phanstiel DH, Westphall MS, Coon JJ (2011) Gas-phase purification enables accurate, multiplexed proteome quantification with isobaric tagging. Nat Methods 8(11):933–935
16. Neilson KA, Keighley T, Pascovici D, Cooke B, Haynes PA (2013) Label-free quantitative shotgun proteomics using normalized spectral abundance factors. In: Ming Z, Timothy V (eds) Proteomics for biomarker discovery. Springer, New York, NY, pp 205–222
17. Pascovici D, Song X, Solomon PS, Winterberg B, Mirzaei M, Goodchild A, Stanley WC, Liu J, Molloy MP (2015) Combining protein ratio p-values as a pragmatic approach to the analysis of Multirun iTRAQ experiments. J Proteome Res 14(2):738–746. doi:10.1021/pr501091e
18. Cox J, Mann M (2008) MaxQuant enables high peptide identification rates, individualized p.p.b.-range mass accuracies and proteome-wide protein quantification. Nat Biotechnol 26(12):1367–1372
19. Oberg AL, Mahoney DW, Eckel-Passow JE, Malone CJ, Wolfinger RD, Hill EG, Cooper LT, Onuma OK, Spiro C, Therneau TM, Bergen HR (2008) Statistical analysis of relative labeled mass spectrometry data from complex samples using ANOVA. J Proteome Res 7(1):225–233. doi:10.1021/pr700734f
20. Herbrich SM, Cole RN, West KP Jr, Schulze K, Yager JD, Groopman JD, Christian P, Wu L, O'Meally RN, May DH, McIntosh MW, Ruczinski I (2013) Statistical inference from multiple iTRAQ experiments without using common reference standards. J Proteome Res 12(2):594–604
21. Meitei NS, Apte A, Biswas A, Pitman M, Samir P, Link AJ (2015) Statistical analysis of multiple iTRAQ/TMT experiments labeled with variable reporter ion tags using ProteoIQ. Poster.

Chapter 6

Unassigned MS/MS Spectra: Who Am I?

Mohashin Pathan, Monisha Samuel, Shivakumar Keerthikumar, and Suresh Mathivanan

Abstract

Recent advances in high resolution tandem mass spectrometry (MS) has resulted in the accumulation of high quality data. Paralleled with these advances in instrumentation, bioinformatics software have been developed to analyze such quality datasets. In spite of these advances, data analysis in mass spectrometry still remains critical for protein identification. In addition, the complexity of the generated MS/MS spectra, unpredictable nature of peptide fragmentation, sequence annotation errors, and posttranslational modifications has impeded the protein identification process. In a typical MS data analysis, about 60% of the MS/MS spectra remains unassigned. While some of these could attribute to the low quality of the MS/MS spectra, a proportion can be classified as high quality. Further analysis may reveal how much of the unassigned MS spectra attribute to search space, sequence annotation errors, mutations, and/or posttranslational modifications. In this chapter, the tools used to identify proteins and ways to assign unassigned tandem MS spectra are discussed.

Key words Mass spectrometry, Peptide, Proteins, Proteomics, Unassigned MS/MS spectra

1 Introduction

Mass spectrometry (MS) based proteomics has been proven to be an indispensable tool for studying perturbation in protein expression [1]. In addition to proteomics research, MS is also used to identify drugs, food contaminants, measure petroleum composition and perform carbon dating [2]. Recently, significant advances have been made to MS instrumentation resulting in tremendous improvements in both resolution and sensitivity in tandem MS data [3, 4]. With the resulting MS/MS spectra, proteins are identified by various search algorithms that predominantly rely on spectral comparison or de novo method. Among these, the most commonly used method in protein identification is based on the spectral comparison of experimental and theoretical MS/MS spectra obtained from sequence databases. Hence, over the years, database search algorithms have remained the gold standard method

Shivakumar Keerthikumar and Suresh Mathivanan (eds.), *Proteome Bioinformatics*, Methods in Molecular Biology, vol. 1549, DOI 10.1007/978-1-4939-6740-7_6, © Springer Science+Business Media LLC 2017

for assigning peptides to MS/MS spectra [5]. However, from thousands of MS/MS spectra, only a fraction of them are mapped to peptides and then to proteins. A majority of the tandem MS spectra remains unassigned. In this chapter, we discuss fundamentals of protein identification methods and limitations of tools required to analyze tandem MS spectra as well as common reasons for these unassigned MS/MS spectra.

2 Methods

There are various search methods used in the identification of peptides and proteins using MS/MS spectra generated from the mass spectrometry. Here, the most commonly used methods such as de novo sequencing and database search are discussed. Depending on the types of fragmentation methods, different fragment ions such as a, b, c, x, y, and z ions are generated. Collision-Induced Dissociation (CID) fragmentation methods generally yields b and y ions, whereas Electron-Transfer Dissociation (ETD) produces mostly c and z ions [6]. Throughout the chapter, b/y ion pairs are discussed for ease of reading.

2.1 De Novo Method

De novo sequencing refers to the identification of amino acid sequence from the tandem MS spectra without any prior knowledge of possible sequences. The methodology is based on the hypothesis that if peptides are fragmented in a predictive manner, MS/MS spectrum will contain the necessary fragment ions to retrieve the entire peptide sequence. De novo sequencing has an advantage of identifying novel peptides and proteins [7]. Basically, de novo sequencing methods takes into account the mass difference between the two adjacent fragment ions to assign the mass of an amino acid residue.

PEAKS is the most commonly used de novo algorithm that identifies peptide sequence among all possible amino acid combinations using dynamic programming algorithm [8]. Interestingly, de novo methods can be used in conjunction with other protein identification algorithms such as the database search method to improve protein identifications. For instance, PEAKS DB is mainly a database search software, but relies on de novo sequencing for better filtration and scoring [9]. Other most commonly used de novo sequencing tools are PepNovo [10] and NovoHMM [11].

2.1.1 Limitations of De Novo Method

MS/MS spectra are complex in nature as the fragmentation patterns can be unpredictable events. In some cases, the charge state of the ions remains ambiguous. Moreover, the complexity of tandem MS spectra is increased significantly when posttranslational modifications are taken into account. For these reasons, it is difficult to predict which of the fragment ions come from b and y ions. Additionally, fragment ions are not identified at certain positions of the peptide resulting in the loss of b/y ion pairs. As de novo sequencing, heavily relies on calculating the mass difference between two adjacent peaks, the complexity of the tandem MS spectra impedes the use of this method in protein identification.

2.2 Database Search

The database search method is the most conventional approach of identification of peptides and proteins from tandem mass spectra using protein sequences. In this approach, protein sequences from the database are digested computationally and hypothetical spectrum is generated for individual peptides. To identify peptide from a MS/MS spectrum, peptides that has total mass equal to the precursor mass of that spectrum are matched to the hypothetically generated spectra from database. As the database search method is based on prior knowledge of the peptide sequence, the b/y ion pairs can be easily identified. Thus, the method overcomes many of the issues faced by de novo sequencing algorithms.

The most commonly used search algorithms such as MASCOT [12], X!Tandem [13] and SEQUEST [14] employ database search strategy for protein identifications. These search programs compares the hypothetical spectra with the observed tandem mass spectrum for each peptide using different scoring methods. The principle behind scoring relies on shared peak count between the theoretical and experimentally observed MS spectrum. Various scoring methods employed by different search programs to assign confidence scores for spectra-peptide match are discussed below.

2.3 Scoring in SEQUEST

SEQUEST is tandem MS data analysis program for the identification of peptides and proteins using database search strategy [14]. In order to score the spectrum against the theoretical spectrum of peptide sequence, SEQUEST uses cross-correlation (XCorr) which is the sum of the peaks that overlap between two spectra. As a measurement of how significant XCorr is, SEQUEST generates autocorrelation (AutoCorr) that measures the alignment of two spectra with a given offset. The ratio of XCorr and average AutoCorr over −75 to +75 Da offset gives a score which is independent of spectral quality and peptide length.

$$R\tau = x[i]\cdot y[i+\tau]$$

$$X\text{corr} = R_0 - \left(\sum_{\tau=-75}^{\tau=75} R_\tau\right)/151$$

where XCorr_1 and XCorr_2 are score for best and second best match, respectively [15].

$$\Delta C_n = \frac{\text{XCorr}_1 - \text{XCorr}_2}{\text{XCorr}_2}$$

ΔC_n is a measure of how good the best match is compared to the second best match.

2.4 Scoring in X! Tandem

X!Tandem is an open source database search tool that matches tandem mass spectrum to hypothetical spectrum generated from protein sequence database. It calculates statistical significance score known as expectation values (*E*-value) for each of the individual spectrum to sequence assignments. X! Tandem's preliminary score is based on the sum of the intensities of matched y and b ions. Hyperscore is then calculated by multiplying preliminary score by factorials of number of b and y ions that are in agreement with experimental spectrum (based on the hypergeometric distribution). X! Tandem assumes that peptides with the highest hyperscore to be the best match. But in cases where the difference between the top hyperscore and the rest is not significantly high, the confidence of identification remains low.

$$y / b\text{Score} = \left(\sum_{i=0}^{n} Ii \times Pi \right)$$

$$\text{HyperScore} = \left(\sum_{i=0}^{n} Ii \times Pi \right) \times Nb! \times Ny!$$

$$\text{where } I_i = \text{intensity of } i\text{th peak and } P_i = \begin{cases} 1 & \text{if } i\text{th peak is predicted} \\ 0 & \text{otherwise} \end{cases}$$

$$N_b = \text{number of b ions } N_y = \text{number of y ions}$$

To measure probability based significance, X!Tandem calculates *E*-value which expresses how unlikely a better hyperscore can occur by random chance [16].

2.5 Scoring in MASCOT

MASCOT is a commercial database search engine for the identification of peptides and proteins using mass spectrometry data. For smaller datasets, MASCOT can be accessed freely using web-based interface at matrix science (http://www.matrixscience.com/) website and can be searched against the default protein sequence databases such as SwissProt and NCBInr. However, for the large scale MS proteomic data analysis, MASCOT license has to be purchased to be used in-house. For fully automated batch search Mascot Daemon utility can be used to submit the request to Mascot server. Though Mascot scoring is mainly based on the MOWSE algorithm and much of the scoring methodology is not published and hence not discussed here.

2.6 Limitations of Database Search

As the database search method mainly relies on known protein sequences in the databases, this approach fails to identify novel peptides and proteins, unidentified mutations and unknown modifications. Even though known modifications can be identified by search algorithms, they are not included in the search parameters most often as they increase the search space (number of candidate peptides) when both modified and unmodified peptides are considered. For instance, in the peptide sequence G <u>I</u> S H <u>V I</u> D, if I and V are considered for a variable modifications, the possible candidates are as follows:

G I S H V I D	G I S H V' I D	G I S H V I' D	G I S H V' I' D
G I' S H V I D	G I' S H V' I D	G I' S H V I' D	G I' S H V' I' D

Due to modification of mere 3 amino acids in this peptide sequence, there are eight (2^3) possible combinations of modifications that needs to be searched. Thus, thousands of peptide sequences increases the search space, computing power and increases false positive hits.

2.7 Unassigned MS/MS Spectra

In spite of the advances in MS instrumentation and software, only a fraction of MS/MS spectra are assigned to peptides [17]. More than 50% of the tandem MS data from not so complex human whole cell lysates are not identified. There are various reasons that can be attributed for the unassigned spectra. Protein identifications using tandem MS relies mainly on known masses of 20 amino acids. There are almost 200 types of modifications [6] and considering all the possible modifications, the search space is computationally intensive. Hence, users may not select many parameters while searching the MS data thereby precluding the identification of those MS/MS spectra that indeed are obtained from modified peptides. Also, the use of database search method inherently accounts for a proportion of these unassigned MS spectra. Novel proteins, exons and sequence variants cannot be identified using this approach as they are not present in the sequence databases. Hence, MS/MS spectra arising from these peptides will not be mapped. Similarly, known mutations and single nucleotide polymorphisms are also responsible for unassigned MS spectra. Lastly, unknown posttranslational modifications and/or unknown chemical modifications during sample processing, can also account for unassigned spectra (*see* Fig. 1).

2.8 Assigning Unassigned

It has to be emphasized that not all the unassigned tandem MS spectra are of good quality; there are proportion of MS/MS spectra that are not high quality and hence can be discarded as noise [18]. High quality unassigned spectra is still worthy of analysis using different search methods or same database search method with increased search parameters. A good quality spectrum is that which has similar statistics in terms of number of peaks, intensity, average distance between peaks in comparison to other assigned spectra [18]. Different types of spectrum metrics such as number of peaks, mean intensity, standard deviation of intensities, number of maximum length of sequence tag identified in the spectrum, number of complementary peak pairs, can be used to assess spectral quality. Good quality spectra can be further analyzed and searched

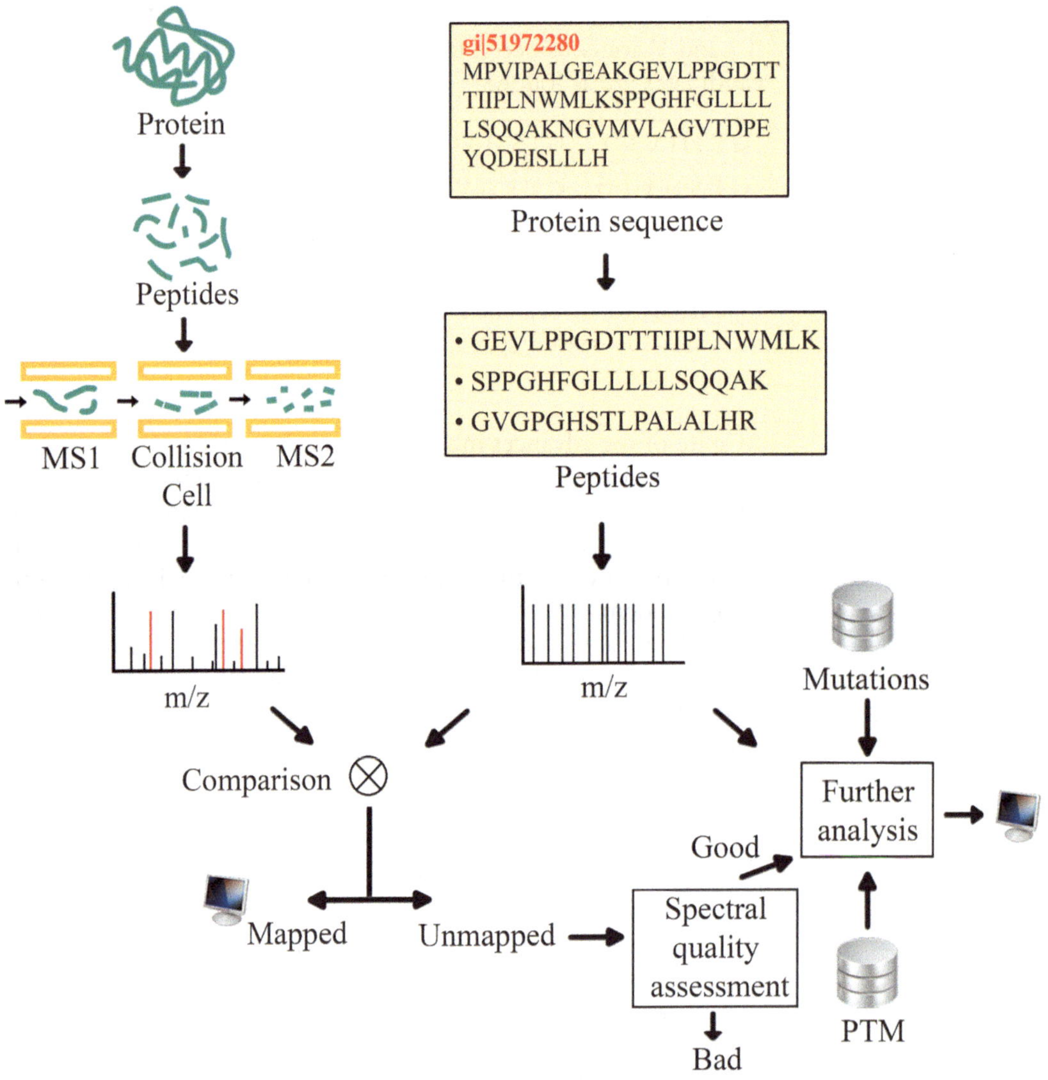

Fig. 1 Typical workflow of database search method extended to identify unassigned tandem MS spectra. Protein samples are reduced, alkylated, and digested by proteases resulting in peptides. The digested peptides are later subjected to LC-MS/MS. The resulting RAW tandem MS data is converted to peak list files by vendor-based and open source software. The resulting peak list files are searched against protein databases to identify as many proteins as possible within the specified FDR. The unmapped MS/MS spectra can be analyzed for the spectral quality and further analyzed incorporating mutations and PTMs

for modifications, mutations, and sequence variants. Proteogenomic approach can further be used to find out possible candidate peptides which are not in protein database by searching genome sequence [19–25] to validate known events as well as to identify novel events such as mutations [17]. Mass tolerant database search [26] can also be used to map unassigned spectra.

3 Conclusions

Analysis tools are required to be developed to separate high and low quality MS/MS spectra. This will allow researchers to only re-search higher quality unassigned spectra and hence will increase peptide assignment. In addition, the contribution of search space and unknown events to the pool of unassigned spectra need to be studied to develop better peptide mapping tools. Though new tools are being developed [3, 4, 27, 28], the mystery of unassigned spectra still remains.

References

1. Aebersold R, Mann M (2003) Mass spectrometry-based proteomics. Nature 422(6928):198–207
2. Maher S, Jjunju FP, Taylor S (2015) Colloquium: 100 years of mass spectrometry: perspectives and future trends. Rev Mod Phys 87(1):113
3. Dorfer V, Pichler P, Stranzl T, Stadlmann J, Taus T, Winkler S, Mechtler K (2014) MS Amanda, a universal identification algorithm optimized for high accuracy tandem mass spectra. J Proteome Res 13(8):3679–3684
4. Wenger CD, Coon JJ (2013) A proteomics search algorithm specifically designed for high-resolution tandem mass spectra. J Proteome Res 12(3):1377–1386
5. Wang P, Wilson SR (2013) Mass spectrometry-based protein identification by integrating de novo sequencing with database searching. BMC Bioinformatics 14(2):1
6. Graves PR, Haystead TA (2002) Molecular biologist's guide to proteomics. Microbiol Mol Biol Rev 66(1):39–63
7. Medzihradszky KF, Chalkley RJ (2015) Lessons in de novo peptide sequencing by tandem mass spectrometry. Mass Spectrom Rev 34(1):43–63
8. Ma B, Zhang K, Hendrie C, Liang C, Li M, Doherty-Kirby A, Lajoie G (2003) PEAKS: powerful software for peptide de novo sequencing by tandem mass spectrometry. Rapid COMMUN Mass Spector 17(20): 2337–2342
9. Zhang J, Xin L, Shan B, Chen W, Xie M, Yuen D, Zhang W, Zhang Z, Lajoie GA, Ma B (2012) PEAKS DB: de novo sequencing assisted database search for sensitive and accurate peptide identification. Molecular & Cellular Proteomics 11 (4):M111. 010587
10. Frank A, Pevzner P (2005) PepNovo: de novo peptide sequencing via probabilistic network modeling. Anal Chem 77(4):964–973
11. Fischer B, Roth V, Roos F, Grossmann J, Baginsky S, Widmayer P, Gruissem W, Buhmann JM (2005) NovoHMM: a hidden Markov model for de novo peptide sequencing. Anal Chem 77(22):7265–7273
12. Cottrell JS, London U (1999) Probability-based protein identification by searching sequence databases using mass spectrometry data. Electrophoresis 20(18):3551–3567
13. Craig R, Beavis RC (2004) TANDEM: matching proteins with tandem mass spectra. Bioinformatics 20(9):1466–1467
14. Eng JK, McCormack AL, Yates JR (1994) An approach to correlate tandem mass spectral data of peptides with amino acid sequences in a protein database. J Am Soc Mass Spectr 5(11):976–989
15. Eng JK, Fischer B, Grossmann J, MacCoss MJ (2008) A fast SEQUEST cross correlation algorithm. J Proteome Res 7(10):4598–4602
16. Ghosh PK (2015) Introduction to protein mass spectrometry. Elsevier Science, Amsterdam
17. Mathivanan S, Ji H, Tauro BJ, Chen Y-S, Simpson RJ (2012) Identifying mutated proteins secreted by colon cancer cell lines using mass spectrometry. J Proteomics 76:141–149
18. Nesvizhskii AI, Roos FF, Grossmann J, Vogelzang M, Eddes JS, Gruissem W, Baginsky S, Aebersold R (2006) Dynamic spectrum quality assessment and iterative computational analysis of shotgun proteomic data toward more efficient identification of post-translational modifications, sequence polymorphisms, and novel peptides. Mol Cell Proteomics 5(4):652–670
19. Romine AO, Bafna V, Smith RD, Pevzner PA Whole proteome analysis of post-translational modifications.

20. Ansong C, Purvine SO, Adkins JN, Lipton MS, Smith RD (2008) Proteogenomics: needs and roles to be filled by proteomics in genome annotation. Brief Funct Genomic Proteomic 7(1):50–62. doi:10.1093/bfgp/eln010
21. Castellana N, Bafna V (2010) Proteogenomics to discover the full coding content of genomes: a computational perspective. J Proteomics 73(11):2124–2135. doi:10.1016/j.jprot.2010.06.007
22. Nesvizhskii AI (2014) Proteogenomics: concepts, applications and computational strategies. Nat Methods 11(11):1114–1125. doi:10.1038/nmeth.3144
23. Keerthikumar S, Gangoda L, Liem M, Fonseka P, Atukorala I, Ozcitti C, Mechler A, Adda CG, Ang CS, Mathivanan S (2015) Proteogenomic analysis reveals exosomes are more oncogenic than ectosomes. Oncotarget 6:15375–15396
24. Prasad TS, Harsha HC, Keerthikumar S, Sekhar NR, Selvan LD, Kumar P, Pinto SM, Muthusamy B, Subbannayya Y, Renuse S, Chaerkady R, Mathur PP, Ravikumar R, Pandey A (2012) Proteogenomic analysis of Candida glabrata using high resolution mass spectrometry. J Proteome Res 11(1):247–260. doi:10.1021/pr200827k
25. Pawar H, Sahasrabuddhe NA, Renuse S, Keerthikumar S, Sharma J, Kumar GS, Venugopal A, Sekhar NR, Kelkar DS, Nemade H, Khobragade SN, Muthusamy B, Kandasamy K, Harsha HC, Chaerkady R, Patole MS, Pandey A (2012) A proteogenomic approach to map the proteome of an unsequenced pathogen – Leishmania donovani. Proteomics 12(6):832–844. doi:10.1002/pmic.201100505
26. Chick JM, Kolippakkam D, Nusinow DP, Zhai B, Rad R, Huttlin EL, Gygi SP (2015) A mass-tolerant database search identifies a large proportion of unassigned spectra in shotgun proteomics as modified peptides. Nat Biotechnol 33(7):743–749
27. Chi H, Chen H, He K, Wu L, Yang B, Sun R-X, Liu J, Zeng W-F, Song C-Q, He S-M (2012) pNovo+: de novo peptide sequencing using complementary HCD and ETD tandem mass spectra. J Proteome Res 12(2):615–625
28. Liu X, Hengel S, Wu S, Tolić N, Pasa-Tolić L, Pevzner PA (2013) Identification of ultra-modified proteins using top-down tandem mass spectra. J Proteome Res 12(12):5830–5838

Chapter 7

Methods to Calculate Spectrum Similarity

Şule Yilmaz, Elien Vandermarliere, and Lennart Martens

Abstract

Scoring functions that assess spectrum similarity play a crucial role in many computational mass spectrometry algorithms. These functions are used to compare an experimentally acquired fragmentation (MS/MS) spectrum against two different types of target MS/MS spectra: either against a theoretical MS/MS spectrum derived from a peptide from a sequence database, or against another, previously acquired MS/MS spectrum. The former is typically encountered in database searching, while the latter is used in spectrum clustering and spectral library searching. The comparison between acquired versus theoretical MS/MS spectra is most commonly performed using cross-correlations or probability derived scoring functions, while the comparison of two acquired MS/MS spectra typically makes use of a normalized dot product, especially in spectrum library search algorithms. In addition to these scoring functions, Pearson's or Spearman's correlation coefficients, mean squared error, or median absolute deviation scores can also be used for the same purpose. Here, we describe and evaluate these scoring functions with regards to their ability to assess spectrum similarity for theoretical versus acquired, and acquired versus acquired spectra.

Key words Mass spectrometry, Scoring functions, Spectrum similarity, Database searching, Spectrum library

1 Introduction

Mass spectrometry (MS) is an essential analytical technique in proteomics [1, 2]. It allows the identification of proteins within a sometimes complex protein mixture. A typical proteomics experiment starts with the digestion of the proteins in the sample into peptides with the aid of proteases. These peptides are subsequently separated via chromatographic techniques and then introduced into a mass spectrometer where a selected peptide is fragmented to yield an MS/MS spectrum [3]. Such an MS/MS spectrum consists of m/z values and the associated intensities for each detected ion. A typical experiment results in the acquisition of (tens of) thousands of MS/MS spectra (Fig. 1a), and these are then assigned to peptides with the aid of computational methods [4–6] (Fig. 1b).

Shivakumar Keerthikumar and Suresh Mathivanan (eds.), *Proteome Bioinformatics*, Methods in Molecular Biology, vol. 1549, DOI 10.1007/978-1-4939-6740-7_7, © Springer Science+Business Media LLC 2017

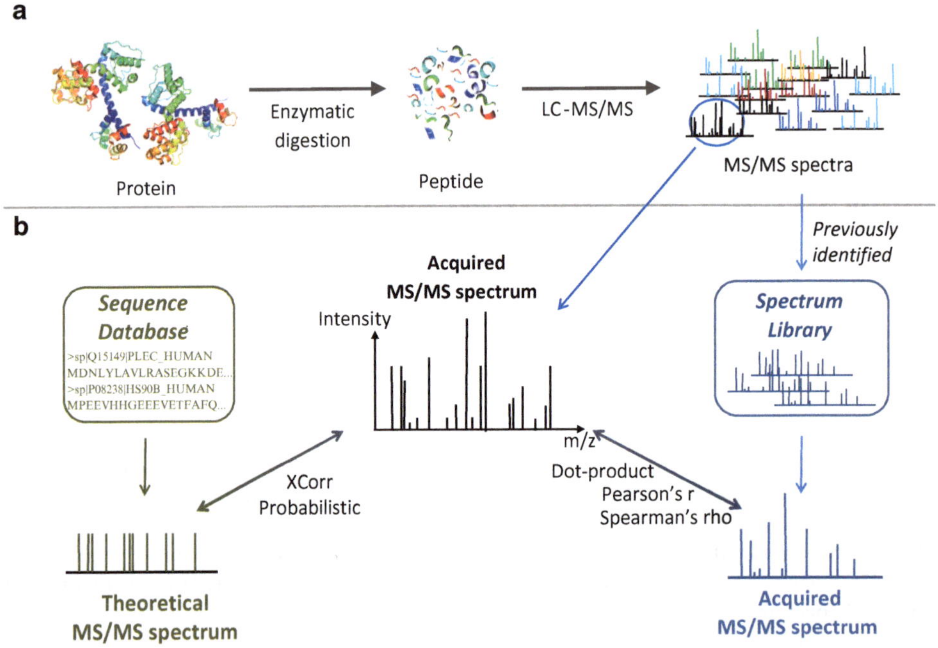

Fig. 1 Overview of a typical proteomics experiment (**a**) and computational methods to match acquired MS/MS spectra (**b**). Acquired spectra can be matched to either a theoretical spectrum generated from a sequence database (e.g., in the database searching approach), or a previously acquired and identified spectrum (e.g., in the spectrum library searching approach). Different scoring functions can be used, including cross-correlation (Xcorr) or probability-based functions for database searching, and dot-product, Pearson's *r* and Spearman's rho scoring functions for spectrum library searches

These computational methods can be divided into two categories (Fig. 1b): those where an acquired MS/MS spectrum is compared against a theoretical MS/MS spectrum, and those where an acquired MS/MS spectrum is compared to another, previously acquired MS/MS spectrum. These comparisons are assigned a score through a function that quantifies the similarities between the spectra [7, 8]. An ideal scoring function should enable the separation of similar from dissimilar MS/MS spectra [9], and do so reasonably quickly [10].

A variety of scoring functions have been implemented for spectral comparison, including cross-correlation [11, 12], normalized dot product [7, 10, 13–15], Pearson's and Spearman's rank correlation coefficients [16, 17], cumulative binomial probability [18, 19], mean squared error, or median absolute deviation. In proteomics, these similarity scores have been applied in database searching [7, 10–13, 18, 19], spectrum library searching [15, 20–22], and spectrum clustering [13, 14].

This review describes MS/MS spectrum similarity scoring functions and their applications in proteomics, and assesses their relative performance on sample data.

2 Methods

2.1 Matching Acquired and Theoretical MS/MS Spectra: Database Searching

Database searching is the method of choice for most researchers to assign peptides to acquired MS/MS spectra [23]. In this method, a suitable protein database is in silico digested to obtain peptides that are subsequently fragmented in silico to yield theoretical MS/MS spectra [24]. These theoretical spectra are in turn matched to the acquired MS/MS spectra by specialized algorithms called search engines [25, 26]. Such a peptide-to-spectrum match (PSM) is scored through a scoring function, which can be either non-probabilistic or probabilistic [27].

2.1.1 Non-probabilistic Scoring Function

The similarity between a theoretical and an acquired MS/MS spectrum can be computed by scoring functions based on matched peaks, such as shared peak count (SPC) [28] that simply reflects the number of matched peaks between paired MS/MS spectra. A more elaborated version is found in the scoring function of Morpheus [29] that relies on the sum of the number of matched peaks and the fraction of matched intensity over total intensity. Another type of non-probabilistic scoring function is the cross-correlation (or *sliding dot product*), computed as

$$R_\tau = \sum_{i=1}^{n} x[i] \cdot y[i+\tau] \tag{1}$$

where n is the number of mass-over-charge (m/z) bins and τ is the relative displacement or sample shifting correction factor. In pairwise MS/MS spectrum similarity, x and y represent the binned versions of the theoretical and the acquired MS/MS spectrum, respectively. $x[i]$ and $y[i]$ reflect the intensity in the ith bin of the respective spectra. τ is the relative displacement between MS/MS spectra; the correction factor.

SEQUEST [12], which is one of the earliest database search algorithms, relies on both an SPC score and a cross-correlation. The first step rounds every acquired peak m/z to the closest integer value. This is followed by the removal of the peaks within 10-u mass window around the precursor ion. Only the 200 most intense peaks are then retained and their intensities are normalized to 100. After this spectrum preprocessing step, the preliminary score (Sp) is calculated as

$$\mathrm{Sp} = \sum i_m n_i (1+\beta)(1+\rho) / n_t \tag{2}$$

where Σi_m is the sum of the matched ion intensities, ni is the number of matched b- and y-ions. $(1+\beta)$ is a scoring part regarding to the ion series continuity and is incremented for every found consecutive theoretical ion (with $\beta = 0.075$). For example, if the number of found consecutive theoretical ions equals to 5, then this value would be $(1 + C\times\beta) = (1 + 5\times 0.075) = 1.375$. $(1+\rho)$ is another scoring component regarding to the existence of immonium ions (with $\rho = 0.15$). SEQUEST assumes five amino acids that yield immonium ions: His, Met, Phe, Tyr, and Trp. In case that an immonium ion is found along with an expected amino acid in the peptide sequence, this value is incremented but in case of the absence of this amino acid in the peptide sequence, this value is decremented. The last scoring component (n_t) is the number of theoretical peaks. Higher Sp scores are expected to be found for true peptide sequences and therefore the top 500 candidate peptide sequences based on Sp scores are selected for the further consideration.

The top 500 candidate peptides are matched against the acquired spectrum by the calculation of cross-correlation-based *final score*. Every theoretical spectrum is constructed of the computed *m/z* values for each fragment ion; also considering their neutral losses. The intensity values of these calculated *m/z* values are assigned to one of three different intensity values, which are [10, 25, 50] chosen based on the fragment ion types. Before starting the comparison with the theoretical spectrum, the original acquired MS/MS spectrum is again preprocessed as follows: peaks within 10-u mass window around the precursor ion are removed. Next, the spectrum is divided into ten intervals and finally, all intensities in each interval are normalized again, but this time to 50. After this preprocessing step, the constructed theoretical spectrum is matched to this preprocessed acquired spectrum with the cross-correlation (xcorr) score calculated as:

$$\text{xcorr} = R_0 - \overline{R_\tau} \tag{3}$$

where xcorr is thus calculated from the difference between the correlation (R) at $\tau = 0$ and the mean of the $R\tau$ correlations with shifted acquired spectra ($\forall \tau \in \mathbb{Z} : -75 < \tau < 75$) (Eq. 3). The calculation of the xcorr is time consuming which prompted researchers to improve the speed of SEQUEST [31], or to develop derived search engines such as Comet [11], Crux [32], and Tide [33].

2.1.2 Probabilistic Scoring Function

Theoretical and acquired MS/MS spectra can also be matched by scoring functions that reflect the probability of finding such a matching score purely by chance. This approach forms the basis of Mascot [34], one of the most popular database search algorithms. The exact details of the Mascot scoring function remain unknown, however, because it is a commercial search engine. The Andromeda search engine [18] is another database search algorithm that uses a probabilistic approach with a published scoring function. Before the comparison starts, if possible, an acquired raw MS/MS spectrum is preprocessed as centroiding, de-isotoping, and charge state deconvolution. After this preprocessing step, the q most intense peaks are remained per 100-u mass window. This filtered spectrum is then compared against a theoretical spectrum generated from a peptide sequence within given precursor tolerance. This theoretical spectrum is constructed such that it always contains singly charged b- and y-ions. However, doubly charged b- and y-ions are also included, in case a precursor ion with a charge state higher than one is observed. Water and ammonium losses are introduced for specific amino acids. After the construction of a theoretical spectrum, an acquired spectrum is matched by the scoring function derived from a cumulative binomial probability as

$$s = -10 \times \log_{10} \sum_{j=k}^{n} \left[\binom{n}{j} (p)^{j} (1-p)^{n-j} \right] \quad (4)$$

where n represents the number of theoretical peaks and k is the number of matched peaks within a given fragment tolerance. p is the probability of finding a single-matched peak by chance and is calculated by dividing the number of the highest-intense peaks (q) by a mass-window size (100-u) (Eq. 4). Because a significant match is such a small value, the logarithm of this computed value is taken, and then multiplied by (−10) to define a score (s).

The same filtered spectrum is also scored against again another theoretical spectrum which now also contains modification-specific losses. In the end, two s scores are computed for the same filtered spectrum and the maximum of these two s scores is selected for this filtered spectrum. As a next step, q is optimized and so a score is computed for every value up to the user defined maximum q value. Finally, the maximum of all computed scores is reported as the final score.

Another search engine that uses a cumulative binomial distribution function is MS-Amanda [19], although there are some differences compared to Andromeda. First of all, MS-Amanda uses n as the number of acquired peaks in a filtered MS/MS spectrum rather than the number of theoretical peaks like in Andromeda. Second, MS-Amanda introduces a direct-weight derived from the peak intensities, which is calculated as the fraction of matched intensity over total intensity on each filtered MS/MS spectrum. In Andromeda, on the other hand, intensities are indirectly used by selecting only the top q most intense peaks per 100-u window during the filtering step. Thirdly, even though the maximum of q was set to 10 for both search engines; the minimum of q was set to 2 and 1 respectively for Andromeda [18] and MS-Amanda [19]. Lastly, MS-Amanda takes into account overlapping peaks in its probability calculation, whereas Andromeda uses a simplified way of a probability calculation which Cox et al. [18] show to work well on high accuracy data.

2.2 Matching Between Acquired MS/MS Spectra: Spectrum Library Searching and Spectrum Clustering

A peptide sequence can also be inferred indirectly from an acquired MS/MS spectrum by comparing it against a previously reliably identified acquired MS/MS spectrum. This approach is called spectrum library searching [35]. A spectrum library is a collection of MS/MS spectra, each of which represents a previously identified peptide [36]. Each such representative MS/MS spectrum is either selected or composed from a set of MS/MS spectra that are all matched to the same peptide. When only a single MS/MS spectrum has been acquired for a given peptide, it is referred to as a singleton [36]. Usually however, multiple MS/MS spectra are reliably matched to a given peptide and a representative MS/MS spectrum is then determined by either choosing the best replicate MS/MS spectrum [22], or by building a consensus MS/MS spectrum [15, 21]. A spectrum library from the best replicate MS/MS spectra can be simply constructed by the input from a user with a list of identified spectra [22]. In the case that multiple spectra for the same peptide were observed, each spectrum is pairwise compared and the average of the computed scores is set as the score for this spectrum; finally the spectrum with the highest average similarity score is selected as the best replicate spectrum [22]. A consensus MS/MS spectrum, on the other hand, is built from multiple similar MS/MS spectra, and this procedure can take many forms. One way to form the consensus spectrum assembly is explained as followed by Lam and coworkers [15]: MS/MS spectra are ranked by signal-to-noise ratio (calculated as the average intensity of the second and sixth highest peaks divided by the median intensity across all peaks). The peaks in these ranked MS/MS spectra are then matched with an adaptive tolerance for the peak m/z, starting from the top ranked spectrum. The consensus spectrum is then created by including only those peaks that were matched in more than 60% of the spectra, and these peaks are then assigned an m/z and intensity value that is calculated as a weighted average, with the weight based on the signal-to-noise ratio [7]. The usage of the consensus spectrum is shown to be more realistic compared to the best-replicate approach [7, 36, 37]. After building such a spectrum library, an acquired MS/MS spectrum can be matched against these representative MS/MS spectra [36]. This is usually performed on the basis of a dot-product score, which is the most commonly used scoring function to match between acquired MS/MS spectra [8, 37].

This spectrum library approach can be considered as an alternative or complementary strategy to a database search, and yields increased speed and sensitivity [7, 15]. The search space in spectrum library searching is built from actual, acquired and identified MS/MS spectra, and each of these spectra is composed of peaks at different m/z with varying intensities. This contrasts sharply with theoretical MS/MS spectra, which are calculated from all putative peptides in a database, and which contain only theoretical fragment ion *m/z* values with (pseudo-)uniform peak intensities. The reduced search space in spectral library searching is an important factor in the speed improvement, while the use of actually observed and realistically varied peaks makes the approach more sensitive. However, the method does require a high quality spectrum library and an effective matching algorithm [36].

A match between acquired MS/MS spectra is also used for other spectral comparison purposes, such as clustering [7, 10, 13] and for finding similarities between data sets within a study [17]. Even though a normalized dot-product is the most commonly used scoring function for these purposes [15, 38–41], Pearson's and Spearman's rank correlation coefficients have also been used to calculate spectrum similarities in some studies [16, 17, 42, 43].

2.2.1 Normalized Dot Product

The *dot product* (or *scalar product*) is a measure that reflects the relative location of two vectors ($x \bullet y$) in space, taking into account their length and direction, and is calculated as (Eq. 5)

$$x \cdot y = \|x\| \, \|y\| \cos\theta = \sum_{i=1}^{n} x_i \times y_i \tag{5}$$

The dot product can be normalized by the product of the *norm* (length) of each vector. The resulting *normalized dot product* is in fact the *cosine* distance between the vectors (Eq. 6):

$$\cos\theta = \frac{\sum_{i=1}^{n} x_i \times y_i}{\sqrt{\sum_{i=1}^{n} x_i^2}\sqrt{\sum_{i=1}^{n} y_i^2}} \tag{6}$$

where the two vectors x and y are derived from data sets with n dimensions: $x = \{x_1, \ldots xn\}$:and $y = \{y_1, \ldots yn\}$. To create such vectors from acquired MS/MS spectra, each spectrum is divided into the same number of n bins, with n either set to a fixed-value [15, 22], or determined based on fragment ion tolerance [17]. Each bin is assigned a certain weight, calculated by summing up all peak intensities in that bin [22], or set to the highest peak intensity in that bin [17]. These binned spectra can now be considered as two vectors of equal dimensionality n, and can be matched against each other by the (normalized) dot product. The normalized dot product ranges between 0 and 1: $\cos\theta = 0$ stems from two orthogonal vectors (MS/MS spectra that have no single peak in common) while $\cos\theta = 1$ is achieved when the two vectors have identical directions (every peak is matched between the MS/MS spectra).

There are several spectrum library search algorithms that are based on a normalized dot product, including SpectraST [15], X!Hunter [21], and BiblioSpec [22]. Probably the most popular library search algorithm, SpectraST [15] primarily relies on consensus

MS/MS spectra, but also offers the option to work with the best-replicate MS/MS spectra.

SpectraST searching starts with filtering out any spectrum (either query spectrum or library spectrum) derived from impurity, which typically has either few peaks (less than six peaks) or negligible signals. This is followed by removing peaks with intensity values lower than the arbitrary set threshold (set to 2.0 in the original paper [15]). Additionally, the intensities of unannotated peaks on the library spectrum are multiplied by 0.2. Subsequently, the square root transformation is applied on the intensities of the remaining peaks. Peaks are then binned into a 1-u window. Later, the normalized dot product is computed between such preprocessed query and library spectra. SpectraST uses two more components to calculate a discriminant scoring function (F) calculated as

$$F = 0.6D + 0.4\Delta D - b \tag{7}$$

where D is the computed normalized dot-product, ΔD is the relative difference between the highest two normalized dot-products. b is a penalty value that is determined according to dot-bias which shows the effect of peaks on the normalized dot-products and there are five different b values for different ranges of the dot-bias values [15].

2.2.2 Correlation Coefficients

Correlation coefficients provide a measure of the linear relationship between two random variables, X and Y. The correlation coefficient for an entire population (ρ) is calculated as

$$\rho(X,Y) = \frac{\mathrm{Cov}(X,Y)}{\sigma_x \sigma_y} \tag{8}$$

However, because population parameters are not known in Eq. 8, these parameters are replaced with sample parameters to calculate the *sample correlation coefficient* (r) (Eq. 9):

$$r_{xy} = \frac{1}{n-1} \sum_{i=1}^{n} \left(\frac{x_i - \bar{x}}{s_x} \right) \left(\frac{y_i - \bar{y}}{s_y} \right) \tag{9}$$

where two data sets with n elements are shown as $x = \{x_1, \ldots, xn\}$ and $y = \{y_1, \ldots, yn\}$ with sx and sy being the standard deviations, respectively. In MS/MS spectrum comparisons, these data sets can be obtained from bin-transformed MS/MS spectra, as explained above.

Correlations are calculated in two different ways based on the type of variables involved. When the variables are continuous and normally distributed, Pearson's product–moment correlation (PPMC, Pearson's correlation, PPC, or Pearson's r in short), rxy, is computed according to Eq. 9. The xi and yi values are equal to the weight of each bin, which reflects the peak intensities in these bins. If the variables are ordinal or not normally distributed, the Spearman's rank correlation (Spearman's rho in short) can be applied as a nonparametric analog of the Pearson's correlation. Spearman's correlation substitutes the weight of the bin by the bin's rank to compute r (Eq. 10). The bins are ranked by ranking the weight of the bin on the preprocessed binned spectrum.

$$r_s = 1 - \frac{6}{n\left(n^2 - 1\right)} \sum_{i=1}^{n} d_i^2 \qquad (10)$$

where $x = \{x_1, \ldots, xn\}$ and $y = \{y_1, \ldots, yn\}$ are the two data sets with n elements each, and with $di = (xi - yi)$ as the difference in rank of the same bin (i) in the two spectra.

The values of both correlation coefficients are confined to $[-1{:}1]$. Positive and negative values show positive and negative correlations, respectively. A correlation of 0 indicates independence of x and y. For Pearson's correlation, $rxy \cong 1$ shows a strong direct linear relationship between x and y; $rxy \cong -1$ shows a strong negative linear relationship between x and y, meaning that x increases with almost the same magnitude as y decreases. For Spearman's correlation, $rs \cong 1$ shows that the ranking for x is very similar as the ranking for y; $rs \cong -1$ shows that the ranking on x is reversed compared to the ranking on y. When comparing MS/MS spectra, $rxy \cong 0$ or $rs \cong 0$ means that the two MS/MS spectra are completely different, $rxy \cong 1$ means that the MS/MS spectra are completely identical, and $rs \cong 1$ means that each MS/MS spectrum has at least the same rank-order of bin intensities.

Further details regarding these scoring functions can be found in [44, 45].

2.2.3 Mean Squared Error

The mean squared error is an estimator based on the differences between two data sets and is computed as

$$\text{MSE} = \frac{1}{n} \sum_{i=1}^{n} \left(x_i - y_i\right)^2 \qquad (11)$$

where $x = \{x_1, \ldots, xn\}$ and $y = \{y_1, \ldots, yn\}$ represent two data sets with n elements each. The differences between the elements in the data sets are squared and then divided by the number of elements. To calculate the more robust median squared error, the median of the squared differences is used instead of the mean. When applied to MS/MS spectra, x and y are derived from binned MS/MS spectra as explained above.

However, compared to the normalized dot product and correlation coefficients, this scoring function has not been commonly used in proteomics to compare two MS/MS spectra.

3 Performance Evaluation of the Different Scoring Functions

3.1 Benchmark Data Sets

The performance of the scoring functions was evaluated on the benchmark data set of the Clinical Proteomic Technology Assessment for Cancer (CPTAC) project of the National Cancer Institute (NCI) [46]. The benchmark data set from Study-6 contains three types of samples: the Sigma-UPS1 48 standard proteins alone, a yeast lysate, and a combination of yeast lysate with Sigma-UPS1 48 standard proteins spiked in at different concentrations. The performance evaluation described here was performed on two MS/MS runs: one run derived from the Sigma-UPS1 sample at 20 fmol/μL (UPS-sample), and the other run from the yeast lysate at 60 ng/μL sample with sigma48 UPS spiked in at 20 fmol/μL (yeast-UPS-sample). The UPS-sample contains 9328 MS/MS spectra, and the yeast-UPS-sample contains 12,089 MS/MS spectra.

3.2 Availability of Codes

An open source Java library [47] enabled to work with the data in computational proteomics. All the Java source codes can be found on https://github.com/compomics/spectrum_similarity.git. Moreover, R source codes to analyze the findings can be found on http://sulesrdiary.blogspot.be/.

3.3 Evaluation of Scoring Functions That Match Acquired Against Theoretical MS/MS Spectra

3.3.1 Spectrum Identification: A Simplified Database Search

The scoring functions that match against theoretical spectra were evaluated with the aid of the method developed by Vaudel et al. [48]. In this study, database searches were performed on *Pyrococcus furiosus* (Pfu) proteins coupled with *Homo sapiens*, Eukaryota, Vertebrata and Mammalia proteins. The results showed that the use of Pfu had a great performance to validate proteomics results. In our case, this method allows us to compare the results from the scoring functions without relying on any database search engine results. We prepared a database that contains the Pfu and the UPS protein sequences (Pfu-UPS) to perform a simplified database search.

The data sets were searched against the concatenated database which consists of the Pfu proteins (4159 protein sequences, downloaded on 13 August 2015, from UniProtKB [49] with taxonomy=2261); the UPS1-UPS2 proteins (50 protein sequences, downloaded on 13 August 2015 from Sigma-Aldrich [50]) and the contaminant proteins (68 protein sequences, downloaded on 13 August 2015 from Global Proteome Machine (GPM) [30]). In silico digestion of this protein database was performed with the aid of DBToolKit (version 4.2.4) [51] with the low peptide mass at 700 Da and the high peptide mass at 2000 Da; and trypsin specificity allowing one miscleavage. This resulted in a total of 132,942 peptides of which 1328 peptides came from the UPS1–UPS2 proteins; 2031 peptides from the contaminant proteins and the remaining peptides originated from Pfu proteins. The database search settings were: 10 ppm as the precursor tolerance and 0.5 Da as the fragment tolerance, with no post-translational modifications selected because the identification of PTMs is difficult and also results in an increased search space and computational time [23, 52, 53].

Any acquired spectrum that contains at least two peaks was compared against a tryptic peptide selected within a given precursor tolerance. The best-ranked peptide (the highest calculated SEQUEST-like and the highest calculated Andromeda-like score) was enlisted for each acquired spectrum, if the peptide was not contaminant-derived. In the case that one spectrum was matched to more than one peptide with the same highest score value, all of these identifications were stored. After finishing the calculation, the identification in agreement between SEQUEST-like and Andromeda-like scores was selected for the further analysis; otherwise the identification for the given spectrum was randomly selected between all of these highest scored identifications.

3.3.2 Theoretical Spectrum Generation

The collision-induced dissociation (CID) fragmentation mode was used during the acquisition of spectra in this study. Therefore, every theoretical spectrum was constructed based on the CID fragmentation mode. A theoretical spectrum contained only b- and y-ions with uniform intensity of 50. If the acquired spectrum in comparison had a precursor charge state higher than one, b- and y-ions were introduced with every charge state varying from charge state one to the precursor charge state. No neutral losses were added to any theoretical spectrum because Degroeve et al. [43], demonstrated that building a model with predicted neutral losses was not as accurate as the model without neutral losses.

3.3.3 SEQUEST-Like Scoring

The final scoring of the original SEQUEST [12] algorithm was implemented as much as possible. However, some steps were altered or skipped and the scoring was therefore called the SEQUEST-like scoring.

The first step on the original SEQUEST and our SEQUEST-like scoring function is processing acquired spectra, followed by the selection of candidate peptides to score. Both scoring functions starts with the precursor peak removal: discarding the peaks in the 10-u window around the m/z-value of the precursor ion. The next step is to keep only the most intense 200 peaks on an acquired spectrum. After keeping the top 200 intense peaks, SEQUEST normalizes intensities to 100 and these spectra are then compared to the list of peptides. These peptides are not only enzyme-specific; instead they are any peptide sequence which may start from any amino acid from the N-terminus onward of any amino acid to the C-terminus of the protein sequence, with the mass within a certain mass tolerance (±3u in their study [12]). These non-enzyme specific peptides were compared by the preliminary score calculation in order to select the 500 highest scored peptide candidates. Instead of selecting such non-enzyme specific peptide-candidates, we selected only tryptic peptides within a precursor tolerance. In addition, we skipped such preliminary score calculation because our aim was to compute the final scoring function on cross-correlation. Besides, such a preliminary score calculation was also not implemented in the SEQUEST-like algorithm Tide [33], which showed to be able to score spectra in a fast and effective way.

After this processing step, an acquired spectrum with at least two peaks was scored against theoretical spectra within given precursor tolerance. Prior to the comparison, each acquired spectrum was divided into ten intervals. For each interval, the most intense peak was set to an intensity of 50 and the intensity of the other peaks was calculated as followed: 50 multiplied with the ratio of the peak intensity over the maximum intensity. Both the processed spectrum and the theoretical spectrum were split into 1u-bins. The binning process started from 75-u before the minimum m/z value of these two spectra to the 75-u after the maximum m/z value of these two spectra. A bin is weighted as the sum of the intensities of the peaks within a corresponding bin. If the m/z of the current peak equals the next bin m/z value, the intensity of this peak was added to the neighboring bin. It is noteworthy to mention that some bins on both binned-spectra may have zero values.

The cross-correlation score was calculated according to the Eq. 3. First, the cross-correlation was calculated between the binned acquired spectrum and the binned theoretical spectrum (R_0). Then, every m/z value of the acquired spectrum was shifted by τ to generate an m/z-shifted binned acquired spectrum, for each integer value of $-75 < \tau < 75$. For every m/z-shifted acquired spectrum, the cross-correlation was calculated and then averaging these calculated values R_τ. This averaged cross-correlation score was subtracted from R_0 to obtain our SEQUEST-like score.

3.3.4 Andromeda-Like Scoring

We have tried to retain the original Andromeda scoring function as much as possible but also here, some adaptations were introduced. Therefore, we call this scoring function the Andromeda-like scoring.

The first adaptation was concerned the theoretical spectrum construction. Andromeda considers always singly charged b- and y-ions and it also introduces doubly charged b- and y-ions in the case that the precursor charge is equal or higher than two. However, in our case, we introduced b- and y-ions with all possible charges; from one to the actual precursor charge (for example, the theoretical spectrum with a triply charged precursor ion has both singly, doubly, and triply charged b- and y- ions). In addition, Andromeda introduces water-, ammonium-, and modification-specific-neutral losses (which can be indeed configured by the user). However, we did not introduce any neutral losses in our theoretical spectrum.

We only considered tryptic peptides and matched acquired spectra against tryptic peptides within a specific precursor tolerance; unlike Andromeda which allows also semi-specific and unspecific enzyme searches.

We did not implement every processing step introduced in Andromeda. In Andromeda, the raw spectrum undergoes several processing steps such as centroiding, de-isotoping, and charge state deconvolution. We however did not include any of these steps; instead, we introduced only the precursor removal step. This step removes peaks that are close to the precursor ion within a given fragment tolerance. We preferred this filtering because another search engine, MS-Amanda, which has a very similar probabilistic scoring function, removes precursor peaks in this way prior to the calculation, and besides also SEQUEST removes precursor peaks.

The calculation of the Andromeda-like score started with dividing a processed-acquired spectrum which contained at least two peaks into 100-u intervals. The topN most intense peaks were selected, with topN varying from 2 to 10. Each processed spectrum with the topN was compared to the theoretical spectrum within the precursor tolerance. First, the number of acquired peaks was matched within a certain fragment tolerance. Based on the probability of finding a peak, computed simply as *topN/100*, the Andromeda-like score was finally calculated with the Eq. 4 for each peptide-to-spectrum match (PSM).

3.4 Results

This evaluation method resulted in the scoring of in total 6265 spectra. The SEQUEST-like scores vary between −7.7*E*+03 and 3.5*E*+04 whereas the Andromeda-like scores vary between 0 and 159.55. A *correct PSM* is a spectrum that both scoring functions matched peptides derived from the UPS proteins (831 PSMs) whereas an *incorrect PSM* is a spectrum that both scoring functions matched peptides derived from the Pfu proteins (5289 PSMs). Some PSMs were considered as *uncertain* because one scoring function assigned a peptide from the UPS proteins and the other scoring function assigned a peptide from the Pfu proteins for the same spectrum. These PSMs were not included in the further analysis (145 out of 6265 PSM, 2.3% were hence excluded). The overall frequency distributions showed that: both distributions were observed to follow a similar trend: The correct PSMs (shaded) tend to give a higher score whereas incorrect PSMs tend (unshaded) to give lower scores (Fig. 2a, b). Moreover, there are more incorrect PSMs compared to the number of correct PSMs, which is not surprising because in our simplified database search, the probability of a random match occurring to a UPS peptide sequence is around 1% (1328 peptides derived from the UPS proteins in 132,942 total putative peptides gives a probability of 9.99*E*−1).

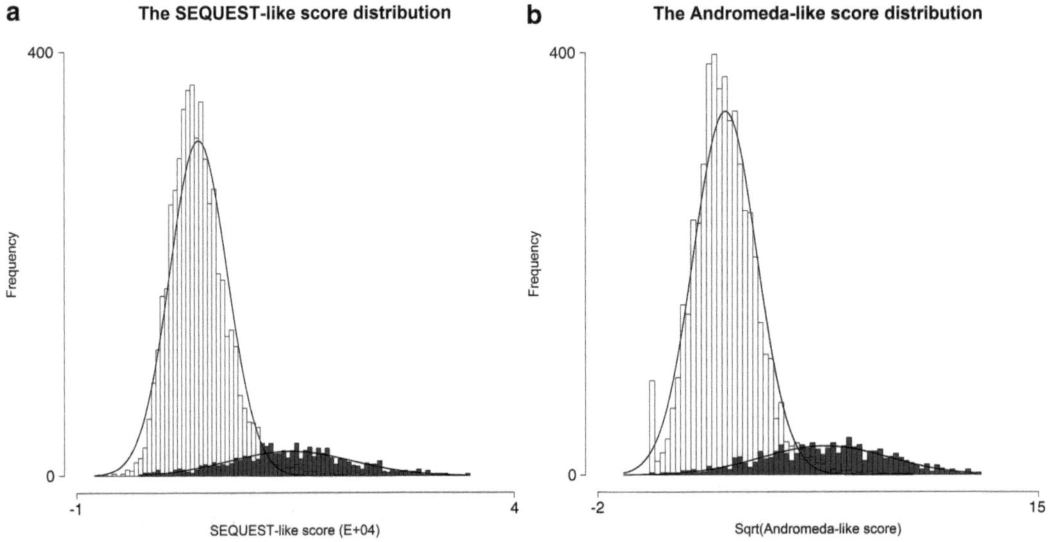

Fig. 2 The frequencies of the correct and incorrect hits for the SEQUEST-like (**a**) and the Andromeda-like (**b**) scores. The frequencies from both scores follow the similar trend. The correct matches (*shaded*) have higher scores than the incorrect matches (*unshaded*)

The two scoring functions present a strong positive relationship (Fig. 3a, PCC = 0.84). In other words, higher SEQUEST-like scores likely correspond to higher Andromeda-like scores. Due to the high number of data points in this plot, it was difficult to analyze how correct or incorrect PSMs correlate to each other. These PSMs were therefore split into two groups: *correct* and *incorrect* PSMs. Correct PSMs again showed a strong positive relationship (PCC = 0.79) with a wide range of scores (Fig. 3b). Incorrect PSMs, however, were not observed to correlate as strong as the correct PSMs (PCC = 0.65) and these incorrect PSMs also had a more narrow and lower score range (Fig. 3c). Furthermore, some matches were assigned to the same peptide by both soring functions, however some other matches were assigned to a different peptide by each soring function. Almost all of the correct PSMs were observed to be assigned to the same peptide (827 PSMs out of 831 PSMs) whereas this percentage dropped to 58.5 % for the incorrect PSMs (3097 PSMs out of 5289 PSMS) (Table 1).

There are two parameters that can influence a scoring function: the peptide length and the precursor charge. Firstly, the peptide length can have an influence on a scoring function because a longer peptide has more fragment ions compared to shorter peptides so scoring functions are not completely independent of the peptide length. Figure 4 shows the effect of a peptide length on the results from only doubly charged identifications (493 correct PSMs versus 3080 incorrect PSMs). The correct PSMs (Fig. 4a, c) tend to have peptides with an amino acid length of 6–14 residues

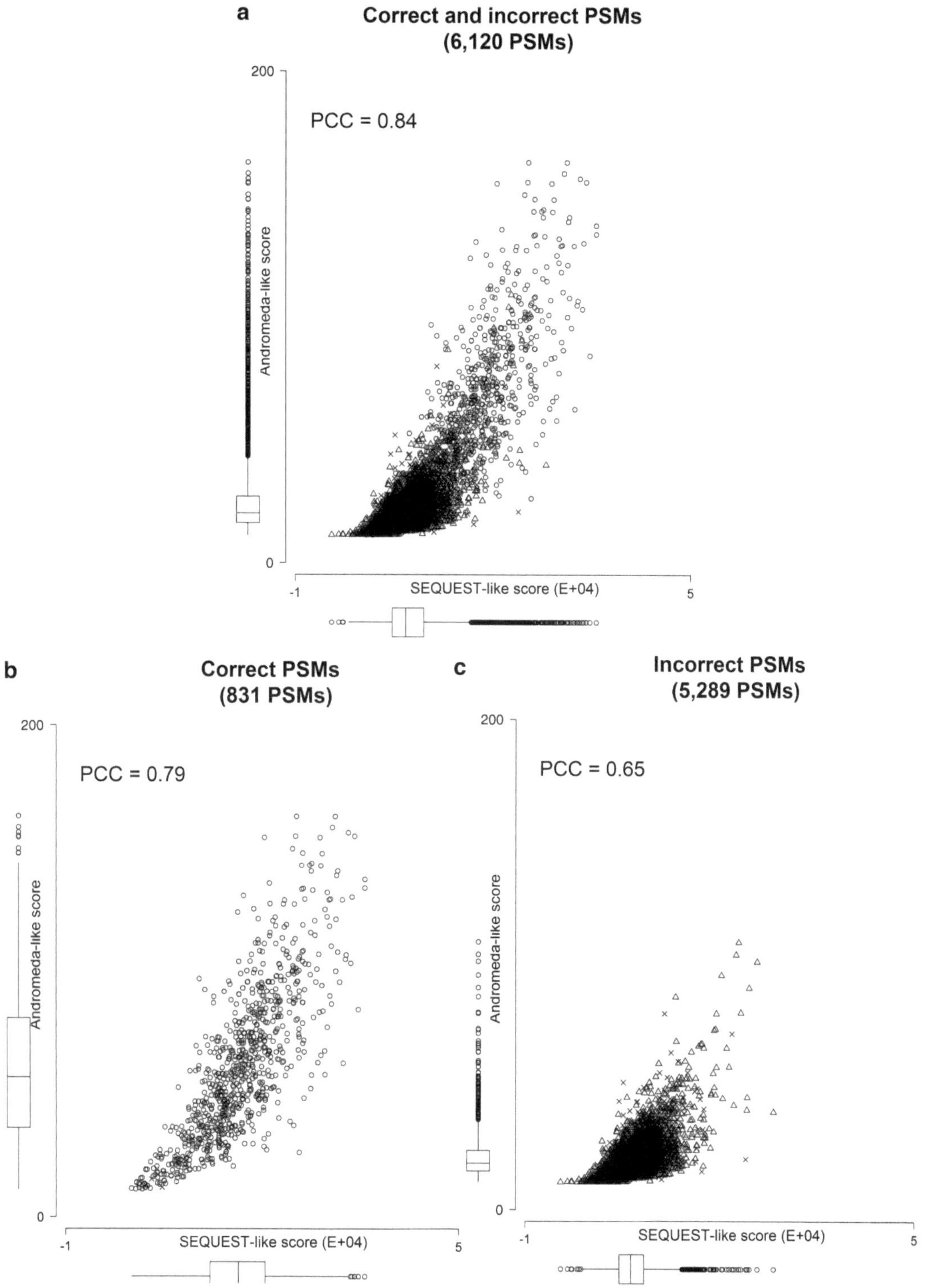

Fig. 3 The score distribution of all correct and incorrect (**a**), only correct (**b**), and only incorrect PSMs (**c**). Any match assigned to the same peptides from the UPS proteins by both scoring functions is shown as an *open circle*. Any match assigned to same peptides from the Pfu protein is shown as an *open triangle*. Any match assigned to different peptides is shown as a cross. In addition to the scatter plot, the score distributions are also seen as *box-plots* on the axis. The SEQUEST-like and the Andromeda-like scoring functions positively correlate to each other with correct hits that tend to have higher scores and incorrect hits that have lower scores

Table 1
The number of PSMs in a simplified database search

	Same peptide sequences	Different peptide sequences	Total assigned peptide sequences
Correct PSMs (only UPS proteins)	827	4	831
Incorrect PSMs (only Pfu proteins)	3097	2192	5289
Uncertain PSMs (both UPS and Pfu proteins)	–	145	145
Total PSMs	3924	2341	6265

A *correct PSM* means that both scoring functions assigned a given spectrum to a peptide derived from a UPS-protein, whereas an *incorrect PSM* means the spectrum was assigned to a peptide from a *Pfu*-protein. An *uncertain PSM* is a spectrum that was assigned to a UPS-protein by one scoring function and a *Pfu*-protein by the other scoring function. The columns show whether the PSM was matched either to the same peptide sequence or to a different peptide sequence

and there is a slight increase on the SEQUEST-like scores of correct PSMs to certain length (around peptide length of 12), whereas the range of Andromeda-like scores are more robust however while the peptide length is becoming longer, Andromeda-like scores decreases more drastically. Because, the increase in peptide length results in finding likely incorrect peaks rather than correct peaks; this affects probabilistic-based scores. Second, longer peptides become less frequent in the database. Secondly, the precursor charge can affect a scoring function (Fig. 5a, b). Acquired peaks can be observed with ion charges from one to the actual precursor charge value depending on the precursor charge. Due to this, theoretical peaks were introduced with different charge states depending on a given precursor charge during theoretical spectrum generation. In our evaluation, every theoretical spectrum was constructed in this manner. Therefore, the peptide matched against the spectrum with a high precursor charge had a noisier theoretical spectrum compared to the theoretical spectrum of a singly charged precursor ion. Because the theoretical spectrum is noisy, the chance of finding an incorrectly matched peak is increased and therefore the probabilistic approach tends to gives lower scores. This can be clearly seen from the results from the Andromeda-like score (Fig. 5b): the scores decrease for correct PSMs (shaded) while the precursor charge is increasing. However, such a drastic change is not observed for the SEQUEST-like score (Fig. 5a).

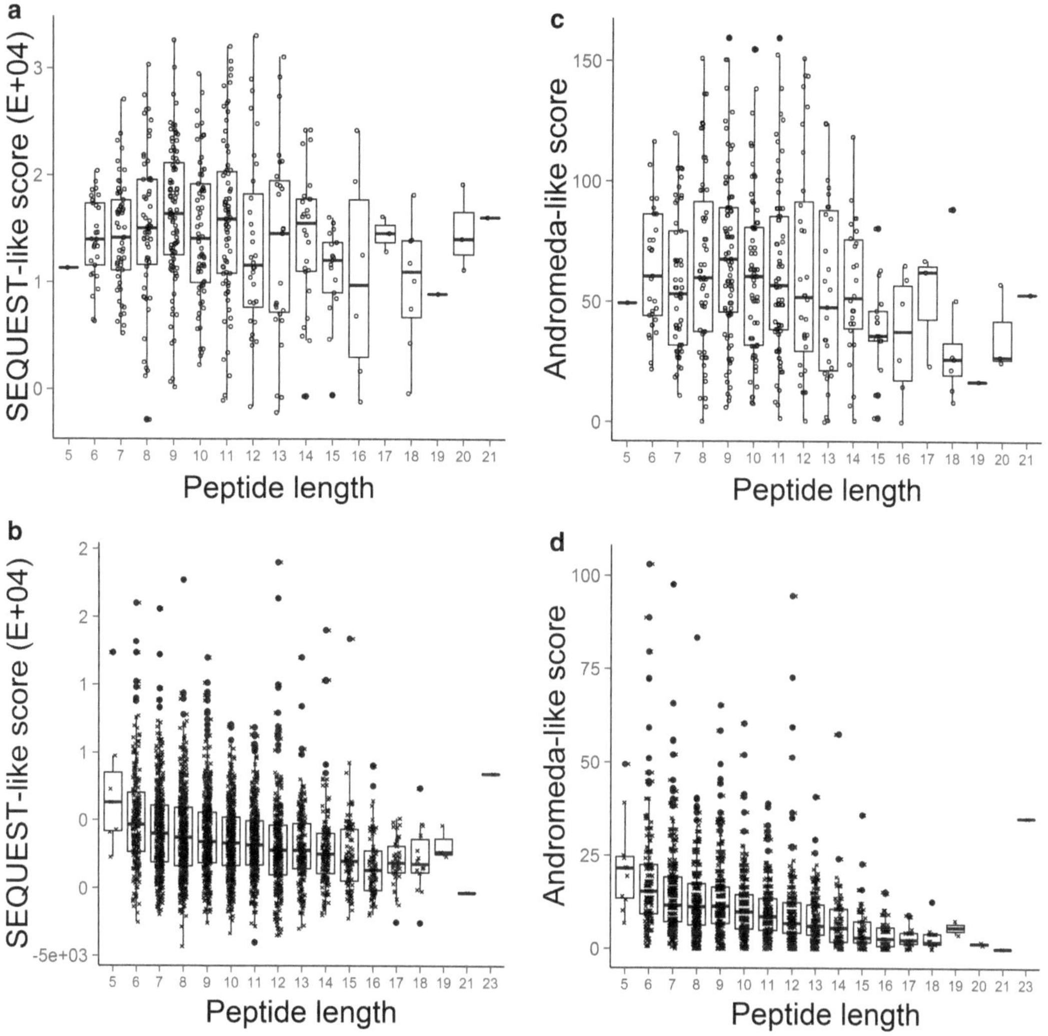

Fig. 4 The effect of the peptide length on the calculated scores. The SEQUEST-like (**a–b**) and Andromeda-like scores (**c–d**) are shown on the *left* and the *right side*. The *x*-axis and the *y*-axis represent respectively the peptide length and the score. The *upper part* shows the correct PSMs and the *lower part* shows the incorrect PSMs

4 Evaluation of Scoring Functions That Match Between Acquired MS/MS Spectra

4.1 Spectrum Identification: Mascot Search

The data sets were searched with Mascot (version 2.4.1) against the database of yeast and UPS1–UPS2 standard proteins (7375 protein sequences including 50 UPS1-UPS2 protein sequences). The database search was performed with the following settings: precursor tolerance of 10 ppm, fragment tolerance of 0.5 Da. Possible precursor charges were set to +2 and +3. The variable modifications selected were acetylation of the N-terminus, carbamidomethylation of cysteine, pyro-Glu formation of N-terminal

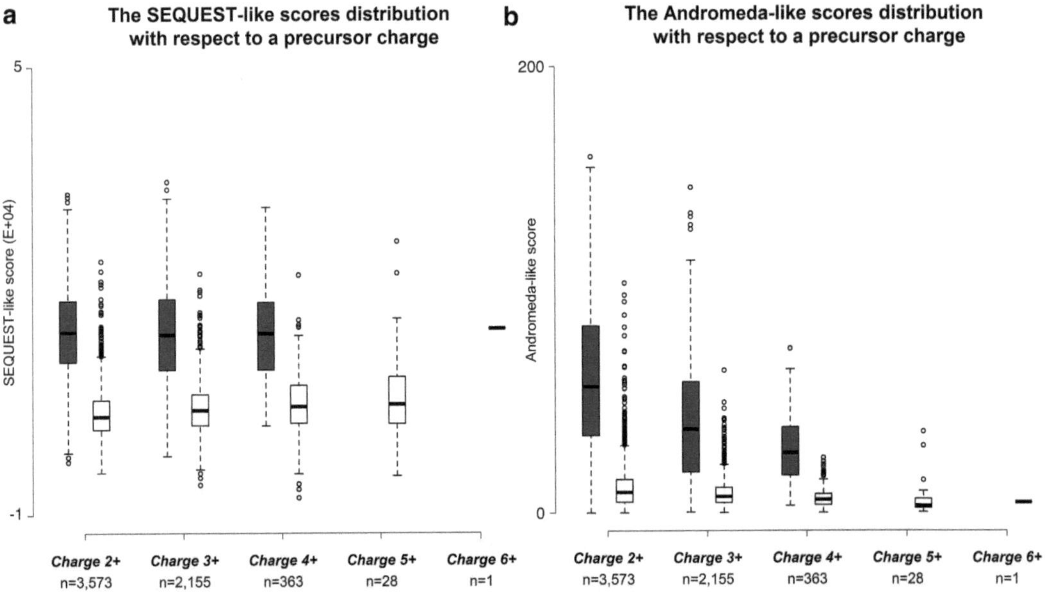

Fig. 5 The effect of the precursor charge on the SEQUEST-like score (**a**) and the Andromeda-like score (**b**). Correct matches are *shaded* whereas incorrect matches are *unshaded*. The increase in the precursor-charge results in the construction of a noisy theoretical spectrum. Therefore, Andromeda-like scores are drastically affected compared to the SEQUEST-like scores

glutamine, pyro-Glu formation of N-terminal glutamate, oxidation of methionine, pyro-carbamidomethylation of N-term cysteine, and no fixed-modification. The enzyme specificity was trypsin and allowing for one missed cleavage. The separate decoy searches were also performed against the shuffled version of the database of yeast and UPS1-UPS2 standard proteins [54].

Mascot identified 359 PSMs on the UPS-data set and 3507 PSMs on the yeast-UPS data set at PEP ≤ 0.05. Two-hundred and seventy four identified peptides were shared between the UPS-data set and the yeast-UPS data sets, and these were identified with the same modification and the same precursor charge.

4.2 Comparison Design

The aim was to perform scoring calculations with different processing settings in order to find UPS-matched MS/MS spectra (identified by Mascot) in the yeast-UPS data set. Therefore, the UPS-matched and the non-UPS-matched MS/MS spectra (both identified by Mascot) from the yeast-UPS data set were compared against the UPS-matched MS/MS spectra (identified by Mascot) from the UPS-data set. This comparison was calculated by four different scoring functions: Pearson's coefficient correlation (Pearson's r), Spearman's coefficient correlation (Spearman's rho), dot-product and mean squared error (MSE).

The initial step was preprocessing spectra and 11 different processing steps were selected (Table 2). Intensities were normalized by either $\log_2$ or the square root transformation (the square root

Table 2
The setting numbers with their associated processing step to perform the comparison against acquired spectra

Setting number	Purpose	Processing
Setting 1	No-processing	None
Setting 2	Intensity normalization	$\log_2$ intensity transformation
Setting 3	Intensity normalization	Square root intensity transformation
Setting 4	Noise filtering	Top50 intense peak
Setting 5	Noise filtering	Top100 intense peak
Setting 6	Noise filtering	Adaptive noise filtering
Setting 7	Noise filtering	Low abundant peaks removal
Setting 8	Precursor peak removal	10-u mass window around precursor ion
Setting 9	Precursor peak removal	Any relevant peaks to the precursor ion
Setting 10	Order of processing steps	Square root transformation—adaptive noise filtering (ordering)
Setting 11	Order of processing steps	Adaptive noise filtering—square root transformation (ordering)

transformation was used for SpectraST [15] and BiblioSpec [22]). Noise peaks were filtered into three options: selecting only the topN intense peaks, applying an adaptive noise filtering [55], and discarding peaks with intensities smaller than 5 % of the maximum intensity. Two values of TopN filtering, TopN = 50 and TopN = 100, were tested because BiblioSpec [22] showed that top50 and top100 performed the best. However, they decided on the top100 intense peak option to enable to work with longer peptides. Lastly, a precursor peak was removed as either discarding peaks around the 10-u mass window or any relevant peaks (as explained on the processing step scoring against theoretical spectrum). After analyzing the results from each setting, the effect of the processing step order was also confirmed by running the comparison by the combination of the best-intensity-transformation and the best-noise-filtering.

The preprocessing step was followed by converting every spectrum into 1-u bins, starting from the minimum and the maximum m/z value in both data sets (min $m/z = 86.05$ was rounded down to 86, and max $m/z = 1994.764$ was rounded up to 1995). For every bin, the intensities within the bin were summed up. Every spectrum could now be scored against another one: Each spectrum in the yeast-UPS data set was matched within a 3-u precursor ion m/z window against the spectra in the UPS-data set. Selected 3-u *m/z* window was in analogy with SpectraST [15] and BiblioSpec [22]. Only spectra with the same charge were scored against another because the different charge state may result in a different spectrum for the same peptide [56]. The best scored-spectrum pair was kept in the list.

This comparison design resulted in 466 UPS-matched and 3026 non-UPS matched MS/MS spectra in the yeast-UPS data set, compared to against 359 UPS-matched MS/MS spectra in the UPS data set. The 466 UPS-matched MS/MS spectra resulted in 406 comparisons and the 3026 non-UPS matched MS/MS spectra resulted in 2359 comparisons, against 351 UPS-matched MS/MS spectra (without any preprocessing step). The 91 UPS-matched MS/MS spectra against the yeast-UPS data set had an exact match to the UPS-data set, which therefore represent *true hits*. Randomly 91 non-UPS-matched MS/MS spectra were selected from the yeast-UPS data set, which represent *false hits*. The scores of these two groups were compared for each scoring function at 11 different processing steps (Table 2). For every score with each setting, the logistic regression model was built with the R package stats and was followed by a ROC curve analysis with the R package pROC [57].

4.3 Results

The best performing score was firstly selected based on the distributions of true and false hits (Fig. 6). Pearson's *r*, Spearman's rho, $\log_{10}$ (Dot-product) and $\log_{10}$ (Mean squared error) scores were therefore computed for the Setting = 1, in which no processing was applied. There is a clear separation between true and false hits only for the Pearson's *r* (Fig. 6). In this setting, Pearson's *r* correlates the best with Spearman's rho but with a correlation value of only 0.55; because these spectra were not processed, therefore the variety of the peak intensity values affects the ranking of Spearman's rho calculation. The highest correlation between any of these paired scores was 0.8 for the $\log_{10}$ (Dot-product) against Spearman. As a side note, we did not include the normalized dot-product in our comparison, because the normalized dot-product behaves very similar to the Pearson's *r* score. A normalized dot-product can therefore separate true hits from false hits, similar to Pearson's *r* score; but as seen in Fig. 6, dot-product only, however, cannot be used for this purpose.

The ROC-curve analysis was further applied to select the best performing scores, the same results without processing (Fig. 7). Pearson's *r* resulted in the highest AUC value (AUC = 0.9902), which shows that Pearson's *r* is definitely able to distinguish true hits from false hits. The second-best performing score function is Spearman's rho with AUC = 0.7727. This AUC value shows that Spearman performs only fair on spectra comparison. MSE was observed to perform the worst (AUC = 0.628) which indicates that this MSE scoring is a poor scoring function to compare spectra without any processing step.

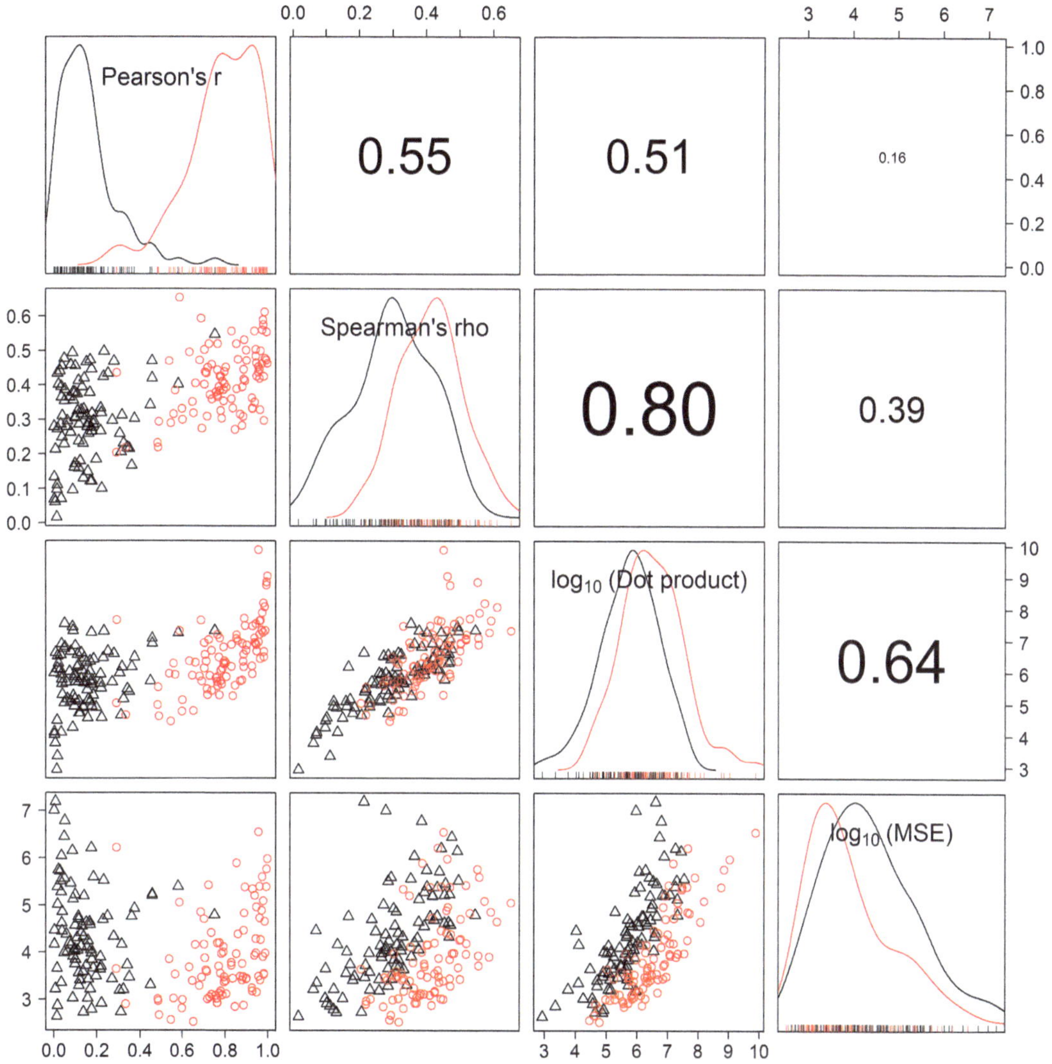

Fig. 6 The score distributions of the true (*open circle* in *red*) and false hits (*open triangle* in *black*) for Pearson's *r*, Spearman's rho, $\log_{10}$ (Dot-product) and $\log_{10}$ (MSE) from the top to the bottom. The *upper panel* shows the correlation values of each pairwise score; the *diagonal* shows the density distributions and the *lower panel* shows the scatter plots of each score (with R-package of car [58])

The ROC-curve analysis was applied to select the best processing settings (Table 3). Pearson's *r* performs rather robust across the different settings (always AUC ≥ 0.9). It is almost always the best performing scoring function except for one setting: top50 filtering in which case Spearman's rho performs the best (AUC = 0.99). Applying a noise filtering improves Spearman's rho scoring; with reaching at least 0.9708. On the other hand, none of the processing steps resulted in a drastic performance improvement for dot-product. AUC for dot-product is always smaller than 0.9, with AUC = 0.75 in average. In addition, MSE performs worse than any

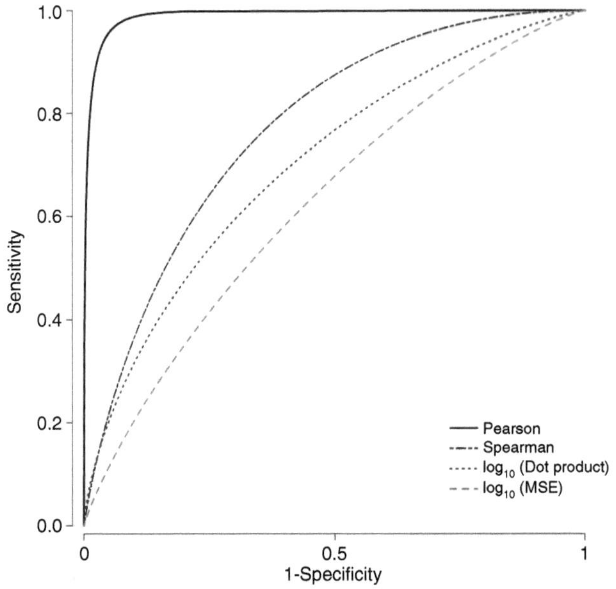

Fig. 7 The ROC curves obtained from the calculation of each score function without any processing (setting = 1). Pearson's *r* (*solid*), Spearman's rho (*two dash*), $\log_{10}$ (Dot-product) (*dotted*), and $\log_{10}$(MSE) (*long dash*) are shown, respectively. Pearson's *r* performs much better than any other scoring function (Pearson's *r* with AUC = 0.9902, the next best performed score of Spearman's rho with AUC = 0.7727)

Table 3
AUC value for every score according to the different process setting

Setting	AUC (Pearson's *r*)	AUC (Spearman's rho)	AUC ($\log_{10}$ (Dot))	AUC ($\log_{10}$(MSE))
Setting 1	0.99017	0.77272	0.7024	0.62801
Setting 2	0.93467	0.79503	0.63851	0.92133
Setting 3	0.9848	0.80848	0.74199	0.76128
Setting 4	0.98898	0.97873	0.7582	0.58639
Setting 5	0.98782	0.99002	0.72844	0.61166
Setting 6	0.99265	0.97084	0.79715	0.62017
Setting 7	0.98815	0.986	0.79857	0.51535
Setting 8	0.98408	0.76814	0.69942	0.63737
Setting 9	0.98539	0.79407	0.73289	0.59258
Setting 10	0.99508	0.99362	0.82926	0.80441
Setting 11	0.99082	0.97801	0.86607	0.75534

other score: AUC = 0.675 in average. However, normalizing intensities into $\log_2$ scale significantly improved this scoring function (AUC = 0.9213). $\log_2$ intensity transformation decreases the magnitude of the intensities more than square root transformation, especially for high values: higher values are affected more compared to small values at $\log_2$ transformation; and the resulting variety between the peak intensities is much less compared to either none or square-root intensity transformation. Lastly, the order of the processing steps was also investigated. Our limited exploration showed that the order matter on three scoring function except Pearson's *r*, which still performed robust with no changes.

5 Conclusion

In this present review, we explain the scoring functions that match acquired spectra against either theoretical or another acquired spectra. In the first half, we compare against theoretical spectra. The most frequently used database search engine functions were implemented as much as possible with some adaptations and these were therefore named as SEQUEST-like and Andromeda-like scoring functions. In the second half, we compare against previously acquired spectra by calculating Pearson's *r*, Spearman's rho, dot-product and MSE with different settings.

SEQUEST-like and Andromeda-like scoring functions were able to separate true hits from false hits; even though these scoring functions use different computational approaches. Both scoring functions are able to find correct matches with assigning the same peptide sequences. Correct matches usually give much higher scores whereas incorrect matches give much lower scores. This illustrates that the combination of the results from different search engines allows the removal of incorrect hits. To achieve this, tools like PeptideShaker [59] and IProphet [60] are available. The fundamental differences between these scoring functions can, however, be seen in the light of comparison against a theoretical spectrum containing more peaks. The increased number of theoretical peaks has a more severe effect on the probabilistic scoring function (Andromeda-like), rather than on the non-probabilistic scoring function (SEQUEST-like). Especially the introduction of theoretical peaks with all possible charges (from one to the precursor charge) increases the chance of matching to a wrong peak. To eliminate this issue, Andromeda [18] introduces singly, for always, and doubly charged theoretical peaks, in case of observing a precursor charge that is higher than one. Moreover, the peptide length also affects scoring functions. Therefore, MaxQuant [61] includes peptide-length for the false discovery rate calculation in peptide identification.

In the comparison against previously acquired spectra, Pearson's r was the most robust approach across the different processing steps. Pearson's *r* is therefore a good alternative to the commonly used normalized dot product. On the other hand, MSE is observed to be a poor scoring function; except one processing setting resulted in a significant improvement. Therefore it is not surprising that this scoring function is not commonly used to compare spectra.

References

1. Domon B, Aebersold R (2006) Mass spectrometry and protein analysis. Science 312:212–217. doi:10.1126/science.1124619
2. Aebersold R, Mann M (2003) Mass spectrometry-based proteomics. Nature 422:198–207. doi:10.1038/nature01511
3. Gevaert K, Van Damme P, Ghesquière B et al (2007) A la carte proteomics with an emphasis on gel-free techniques. Proteomics 7:2698–2718. doi:10.1002/pmic.200700114
4. Eidhammer I, Flikka K, Martens L, Mikalsen S-O (2007) Computational methods for mass spectrometry proteomics. John Wiley & Sons, Ltd, West Sussex
5. Käll L, Vitek O (2011) Computational mass spectrometry-based proteomics. PLoS Comput Biol 7:e1002277. doi:10.1371/journal.pcbi.1002277
6. Xu C, Ma B (2006) Software for computational peptide identification from MS-MS data. Drug Discov Today 11:595–600
7. Lam H, Deutsch EW, Eddes JS et al (2008) Building consensus spectral libraries for peptide identification in proteomics. Nat Methods 5:873–875. doi:10.1038/nmeth.1254
8. Shao W, Zhu K, Lam H (2013) Refining similarity scoring to enable decoy-free validation in spectral library searching. Proteomics 13:3273–3283. doi:10.1002/pmic.201300232
9. Yen C-Y, Houel S, Ahn NG, Old WM (2011) Spectrum-to-spectrum searching using a proteome-wide spectral library. Mol Cell Proteomics 10:M111.007666. doi:10.1074/mcp.M111.007666
10. Kim S, Pevzner P (2014) MS-GF+ makes progress towards a universal database search tool for proteomics. Nat Commun 5:5277. doi:10.1038/ncomms6277
11. Eng JK, Jahan T, Hoopmann MR (2013) Comet: an open-source MS/MS sequence database search tool. Proteomics 13:22–24. doi:10.1002/pmic.201200439
12. Eng JK, McCormack AL, Yates JR (1994) An approach to correlate tandem mass spectral data of peptides with amino acid sequences in a protein database. J Am Soc Mass Spectrom 5:976–989
13. Tabb DL, MacCoss MJ, Wu CC et al (2003) Similarity among tandem mass spectra from proteomic experiments: detection, significance, and utility. Anal Chem 75:2470–2477. doi:10.1021/ac026424o
14. Griss J, Foster JM, Hermjakob H, Vizcaíno JA (2013) PRIDE cluster: building a consensus of proteomics data. Nat Methods 10:95–96. doi:10.1038/nmeth.2343
15. Lam H, Deutsch EW, Eddes JS et al (2007) Development and validation of a spectral library searching method for peptide identification from MS/MS. Proteomics 7:655–667
16. Frank AM (2009) Predicting intensity ranks of peptide fragment ions. J Proteome Res 8:2226–2240. doi:10.1021/pr800677f
17. Li S, Arnold RJ, Tang H, Radivojac P (2011) On the accuracy and limits of peptide fragmentation spectrum prediction. Anal Chem 83:790–796. doi:10.1021/ac102272r
18. Cox J, Neuhauser N, Michalski A et al (2011) Andromeda: a peptide search engine integrated into the MaxQuant environment. J Proteome Res 10:1794–1805. doi:10.1021/pr101065j
19. Dorfer V, Pichler P, Stranzl T et al (2014) MS Amanda, a universal identification algorithm optimized for high accuracy tandem mass spectra. J Proteome Res 13:3679–3684
20. Yates JR, Morgan SF, Gatlin CL et al (1998) Method to compare collision-induced dissociation spectra of peptides: potential for library searching and subtractive analysis. Anal Chem 70:3557–3565. doi:10.1021/ac980122y
21. Craig R, Cortens JC, Fenyo D, Beavis RC (2006) Using annotated peptide mass spectrum libraries for protein identification. J Proteome Res 5:1843–1849. doi:10.1021/pr0602085

22. Frewen BE, Merrihew GE, Wu CC et al (2006) Analysis of peptide MS/MS spectra from large-scale proteomics experiments using spectrum libraries. Anal Chem 78:5678–5684. doi:10.1021/ac060279n
23. Vaudel M, Sickmann A, Martens L (2012) Current methods for global proteome identification. Expert Rev Proteomics 9:519–532. doi:10.1586/epr.12.51
24. Steen H, Mann M (2004) The ABC's (and XYZ's) of peptide sequencing. Nat Rev 5:699–711. doi:10.1038/nrm1468
25. Nesvizhskii A (2007) Protein identification by tandem mass spectrometry and sequence database searching. Mass Spectr Data Anal Proteomics 367:87–119
26. Matthiesen R (2007) Methods, algorithms and tools in computational proteomics: a practical point of view. Proteomics 7:2815–2832. doi:10.1002/pmic.200700116
27. Eidhammer I, Flikka K, Martens L, Mikalsen S-O (2007) Spectral comparisons. Computational methods for mass spectrometry proteomics. John Wiley & Sons, Ltd., West Sussex, pp 159–178
28. Kapp E, Schütz F (2007) Overview of tandem mass spectrometry (MS/MS) database search algorithms. Curr Protoc Protein Sci 25(2):1–19
29. Wenger CD, Coon JJ (2013) A proteomics search algorithm specifically designed for high-resolution tandem mass spectra. J Proteome Res 12:1377–1386
30. GPM The cRAP FASTA file. ftp://ftp.thegpm.org/fasta/cRAP/. Accessed 13 Aug 2015
31. Eng JK, Fischer B, Grossmann J, Maccoss MJ (2008) A fast SEQUEST cross correlation algorithm. J Proteome Res 7:4598–4602. doi:10.1021/pr800420s
32. Park CY, Klammer AA, Käll L et al (2008) Rapid and accurate peptide identification from tandem mass spectra. J Proteome Res 7:3022–3027
33. Diament BJ, Noble WS (2011) Faster SEQUEST searching for peptide identification from tandem mass spectra. J Proteome Res 10:3871–3879. doi:10.1021/pr101196n
34. Perkins DN, Pappin DJC, Creasy DM, Cottrell JS (1999) Probability-based protein identification by searching sequence databases using mass spectrometry data. Electrophoresis 20:3551–3567
35. Hu Y, Li Y, Lam H (2011) A semi-empirical approach for predicting unobserved peptide MS/MS spectra from spectral libraries. Proteomics 11:4702–4711. doi:10.1002/pmic.201100316
36. Lam H (2011) Building and searching tandem mass spectral libraries for peptide identification. Mol Cell Proteomics 10:R111.008565
37. Flikka K, Meukens J, Helsens K et al (2007) Implementation and application of a versatile clustering tool for tandem mass spectrometry data. Proteomics 7:3245–3258. doi:10.1002/pmic.200700160
38. Beer I, Barnea E, Ziv T, Admon A (2004) Improving large-scale proteomics by clustering of mass spectrometry data. Proteomics 4:950–960. doi:10.1002/pmic.200300652
39. Tabb DL, Thompson MR, Khalsa-Moyers G et al (2005) MS2Grouper: group assessment and synthetic replacement of duplicate proteomic tandem mass spectra. J Am Soc Mass Spectrom 16:1250–1261. doi:10.1016/j.jasms.2005.04.010
40. Wan KX, Vidavsky I, Gross ML (2002) Comparing similar spectra: from similarity index to spectral contrast angle. J Am Soc Mass Spectrom 13:85–88. doi:10.1016/S1044-0305(01)00327-0
41. Stein SE, Scott DR (1994) Optimization and testing of mass spectral library search algorithms for compound identification. J Am Soc Mass Spectrom 5:859–866. doi:10.1016/1044-0305(94)87009-8
42. Degroeve S, Maddelein D, Martens L (2015) MS2PIP prediction server: compute and visualize MS2 peak intensity predictions for CID and HCD fragmentation. Nucleic Acids Res 43:W326–W330. doi:10.1093/nar/gkv542
43. Degroeve S, Martens L (2013) MS2PIP: a tool for MS/MS peak intensity prediction. Bioinformatics. doi:10.1093/bioinformatics/btt544
44. Rosner B (2010) Regression and correlation methods., Fundamentals of Biostatistics
45. Eidhammer I, Barsnes H, Eide GE, Martens L (2013) Appendix A: statistics. Computational and statistical methods for protein quantification by mass spectrometry. John Wiley & Sons, Ltd, West Sussex
46. Paulovich AG, Billheimer D, Ham A-JL et al (2010) Interlaboratory study characterizing a yeast performance standard for benchmarking LC-MS platform performance. Mol Cell Proteomics 9:242–254. doi:10.1074/mcp.M900222-MCP200
47. Barsnes H, Vaudel M, Colaert N et al (2011) compomics-utilities: an open-source Java library for computational proteomics. BMC Bioinform 12:70. doi:10.1186/1471-2105-12-70
48. Vaudel M, Burkhart JM, Breiter D et al (2012) A complex standard for protein identification,

designed by evolution. J Proteome Res 11: 5065–5071. doi:10.1021/pr300055q

49. The Uniprot Consortium (2015) UniProt: a hub for protein information. Nucleic Acids Res 43:D204–D212. doi:10.1093/nar/gku989
50. Sigma-Aldrich The UPS FASTA File. http://www.sigmaaldrich.com/content/dam/sigma-aldrich/life-science/proteomics-and-protein/ups1-ups2-sequences.fasta. Accessed 13 Aug 2015
51. Martens L, Vandekerckhove J, Gevaert K (2005) DBToolkit: processing protein databases for peptide-centric proteomics. Bioinformatics 21:3584–3585. doi:10.1093/bioinformatics/bti588
52. Parker CE, Mocanu V, Mocanu M et al (2010) Mass spectrometry for post-translational modifications. Neuroproteomics 2010:PMID: 21882444
53. Allmer J (2010) Existing bioinformatics tools for the quantitation of post-translational modifications. Amino Acids. doi:10.1007/s00726-010-0614-3
54. Gonnelli G, Stock M, Verwaeren J et al (2015) A decoy-free approach to the identification of peptides. J Proteome Res 14:1792–1798. doi:10.1021/pr501164r
55. Hulstaert N, Reisinger F, Rameseder J et al (2013) Pride-asap: automatic fragment ion annotation of identified PRIDE spectra. J Proteomics 95:89–92. doi:10.1016/j.jprot.2013.04.011
56. Liu J, Bell AW, Bergeron JJM et al (2007) Methods for peptide identification by spectral comparison. Proteome Sci 5:3. doi:10.1186/1477-5956-5-3
57. Robin X, Turck N, Hainard A et al (2011) pROC: an open-source package for R and S+ to analyze and compare ROC curves. BMC Bioinform 12:77. doi:10.1186/1471-2105-12-77
58. Fox J, Weisberg S (2011) An R companion to applied regression, 2nd edn. Sage, Thousand Oaks, CA
59. Vaudel M, Burkhart JM, Zahedi RP et al (2015) PeptideShaker enables reanalysis of MS-derived proteomics data sets. Nat Biotechnol 33:22–24. doi:10.1038/nbt.3109
60. Shteynberg D, Nesvizhskii I, Moritz RL, Deutsch EW (2013) Combining results of multiple search engines in proteomics. Mol Cell Proteomics 12:2383–2393. doi:10.1074/mcp.R113.027797
61. Cox J, Mann M (2008) MaxQuant enables high peptide identification rates, individualized p.p.b.-range mass accuracies and proteome-wide protein quantification. Nat Biotechnol 26:1367–1372. doi:10.1038/nbt.1511

Chapter 8

Proteotypic Peptides and Their Applications

Shivakumar Keerthikumar and Suresh Mathivanan

Abstract

Recent advances in mass spectrometry based proteomic techniques and publicly available large proteomic repositories are being exploited to characterize the proteome of multiple organisms. While humongous amount of proteomic data is being acquired and analyzed, many biological questions still remain unanswered. Proteotypic peptides which uniquely represent target proteins or a protein isoform are used as an alternative strategy for protein identification in the field of immunological methods and targeted proteomic techniques. Using different computational approaches, resources and techniques used in the identification of proteotypic peptides of target proteins is discussed here.

Key words Targeted proteomics, Selected reaction monitoring, Biomarkers, Databases, Bioinformatics

1 Introduction

Using different strategies, peptide sequences are matched to tandem mass spectrometry-derived spectra but most of these methods are time consuming due to the large size of the background databases. Besides, detection and quantification of some biomolecules are often below the detection limits of shotgun mass spectrometry based proteomics. As a result, targeted proteomic techniques are largely gaining importance mainly in the field of biomarker validation in blood plasma [1–5]. One such approach is selected reaction monitoring (SRM) also known as multiple reaction monitoring (MRM) in which proteomic experiment is performed by selectively monitoring the peptides of protein sequences with known m/z values (precursor ions) fragmenting through collision induced dissociation and monitoring specific preselected daughter/fragment ions (product ions). Such peptides that uniquely represent targeted proteins or protein isoform are known as proteotypic peptides [1, 6, 7]. Identifications of proteotypic peptides clearly represent the presence of that protein in the sample under investigation and this would further improve the speed and accuracy of protein identifications. Over the last decade,

Shivakumar Keerthikumar and Suresh Mathivanan (eds.), *Proteome Bioinformatics*, Methods in Molecular Biology, vol. 1549, DOI 10.1007/978-1-4939-6740-7_8, © Springer Science+Business Media LLC 2017

proteotypic peptides in mass spectrometry based protein quantification are used as an alternative strategy to antibody-based detection methods. Unavailability of specific antibodies, single analyte testing, and the lack of epitope specificity are known to be the main disadvantages of immunological methods which can be avoided by using proteotypic peptides of target proteins in high throughput analysis [1, 8, 9]. Due to high importance of these proteotypic peptides in the field of clinical proteomics many resources and bioinformatics tools have been developed to predict and store proteotypic peptides to further aid the proteomics research community.

Here, we review the targeted proteomics techniques employed using these proteotypic peptides, proteomics repositories, computational tools and algorithms developed for dissemination and identification of proteotypic peptides in the field of proteomics research.

2 Targeted Proteomics

Identification of low abundant proteins of interest still remains a major drawback in proteomics analysis. As a result targeted biological questions remains unanswered in spite of acquiring several magnitudes of data and its analyses. Emerging targeted proteomics approach seems to be the major solution to obtain quantitative information about the targeted proteins of interest [10]. The techniques and computational approaches used in the targeted proteomics are discussed below.

2.1 Selected Reaction Monitoring

Selected reaction monitoring (SRM) also known as multiple reaction monitoring (MRM) is a targeted technique emerging in the field of proteomics as complement to untargeted shotgun approach. SRM utilizes unique capabilities of triple quadruple (QQQ) mass spectrometers to act as mass filters to selectively monitor a specific analyte molecular ion and one or several fragment ions generated from the analyte by fragmentation methods. Combination of such precursor–fragment ion pairs, termed SRM transitions, can be sequentially and repeatedly measured at a periodicity that is fast compared to the analyte's elution, yielding chromatographic peaks for each transition that allow for the concurrent quantification of multiple analytes [7, 11–14].

2.2 Identification of Proteotypic Peptides

The selection of target proteins and peptides list is one of the major prerequisite of targeted proteomics workflow. The target protein selection entirely depends on the specific biological question that needs to be answered. Besides, identifying proteotypic peptides that uniquely represent target proteins is one of the main challenges of targeted proteomics. Proteotypic peptides with lengths of

~7–23 amino acids, analyzed by triple quadruple in multiple reaction monitoring are usually selected in the targeted proteomics [8]. In general, many physicochemical properties of the peptides are considered for predicting the proteotypic peptides of high signal response. Besides, short hydrophilic and long hydrophobic peptides are avoided, whereas fully tryptic peptides with an average length of ~10 amino acids, devoid of residues prone to artifactual or posttranslational modifications are targeted [7].

Currently, there are few computational online resources such as Global Proteome Machine Database (GPMDB), PeptideAtlas and PRIDE available for the identification of proteotypic peptides for targeted proteins. These proteomic resources are continuously being used further to develop algorithms and bioinformatics tools for the prediction of proteotypic peptides.

2.2.1 Computational Proteomic Resources for the Identification of Proteotypic Peptides

With the massive increase in the application of mass spectrometry based proteomics research new mass spectrometers are being introduced into the proteomics field rapidly which generates humongous amount of tandem mass spectrometry (MS/MS) based proteomics data. Initiative from the proteomics community to collate these proteomics data generated from different experimental strategies using different mass spectrometer instruments resulted in many proteomics repositories for storage and dissemination of proteomics data to aid proteomics research community. To normalize the data generated from different experimental research groups majority of these repositories have developed in-house proteomics pipelines for the identification of significant peptides and proteins. The most commonly used proteomics repositories which can be exploited further to generate proteotypic peptides of target proteins are discussed below.

Global Proteome Machine Database (GPMDB)

The Global Proteome Machine Database (http://www.thegpm.org/) is an open source mass spectrometry based proteomic repository, publicly available for the scientific community developed by Beavis informatics. The GPMDB periodically checks all the public proteomic repositories, downloads and reanalyzes the proteomic data using X! Tandem search engine. The resultant peptide and protein list after passing through the stringent automated quality test are stored into the backend database along with relevant metadata. Further, the results can be either viewed in the GPM website or downloaded through ftp or other interfaces. Besides, the users can also submit their spectra files in different formats such as .mgf, mzXML, pkl, mzData, dta, and common (for only big and compressed files) to GPM via 'Search Data' option available in the website. The most frequently checked public repositories for the suitable new proteomic data for reanalysis includes Proteome Xchange/PRIDE, PeptideAtlas/PASSEL, MassIVE (http://www.massive.ucsd.edu/), Proteomics DB, The Chorus Project (http://chorusproject.org/), and iProX (http://www.iprox.org/).

Recently, at the time of writing this chapter, the GPMDB released an updated version of the GPM Personal Edition-Fury to replace the old venerable Cyclone version and upgraded to the latest version of X! Tandem (Version 2015.12.15, Vengeance) which features speedy PTMs assignments. In addition, the human and mouse protein identification information in GPMDB has been summarized into a collection of spreadsheets known as GPMDB Guide to Human Proteome (GHP) and GPMDB Guide to Mouse Proteome (GMP) respectively. This guide contains information organized into separate spreadsheets for each chromosome as well as mitochondrial DNA and made available for download at ftp://ftp.thegpm.org/projects/annotation/human_protein_guide/ and ftp://ftp.thegpm.org/projects/annotation/proteome_protein_guide/. The GPMDB also hosts spectral search engine called X! Hunter (http://xhunter.thegpm.org/) and proteotypic profiler called X! P3 (http://p3.thegpm.org/) for the analysis of proteomics data. The X! P3 (Proteotypic Peptide Profiler) is known to be the first publicly available search engine for proteotypic peptide profiling built using the X! Tandem refinement idea and the open source X! Tandem code. In order to find the best peptides that uniquely represents target proteins, the X! P3 utilizes proteomics data stored in the GPMDB [15].

PeptideAtlas

The PeptideAtlas (http://www.peptideatlas.org/) database is another freely available mass spectrometry derived proteomic data repository developed at Institute of Systems Biology, Seattle, USA. The PeptideAtlas accepts only spectra files either in the form of RAW, mzML or mzXML format and limited metadata. Once submitted, the raw spectra files are processed using standardized data processing pipeline known as Trans Proteomics Pipeline (TPP) [16] and stored in the SBEAMS (Systems Biology Experiment Analysis Management System)-Proteomics module. Further, the identified highly significant scoring peptide sequences are mapped to their respective genome sequence representing species/sample specific build [17, 18]. Currently, the PeptideAtlas has 19 organism specific build which includes many model organisms such as human, yeast, *C. elegans*, mouse, Drosophila, rat, horse, and zebrafish, for important sample groups such as plasma, brain liver, lung, colon cancer, heart, kidney, and urine.

The PeptideAtlas, similar to the PRIDE archive system, one of the founding members of PX consortium implements standardization of the mass spectrometry proteomics data and automate the sharing of proteomic data across different repositories. Another important feature of the PeptideAtlas is investigation of proteotypic peptides. Currently, users can search proteotypic peptides from three different organisms such as human, mouse, and yeast. Identification of such high scoring peptides would further serve as most possible targets for Selected Reaction Monitoring (SRM) approach [19]. The PeptideAtlas SRM Experiment Library (PASSEL) is a component of the PeptideAtlas project that is designed to enable submission, dissemination, and reuse of SRM experimental results from analysis of biological samples. The raw data submitted via PASSEL are automatically processed and stored into the database which can be further downloaded or accessed via web interface [20].

Further, the distinct peptides and its associated proteins identified from the users submitted raw data files using TPP tool can be further depicted graphically in Cytoscape [21] plugin implemented in the PeptideAtlas. Overall, the PeptideAtlas depicts the normalized outlook of the user submitted data which further aid in genome annotation of different organisms using mass spectrometry derived proteomic data.

2.2.2 Computational Predictions and Identification of Proteotypic Peptides

While many proteomics repositories were developed for the storage and dissemination of mass spectrometry based proteomics data, many proteomics tools and algorithms were developed to exploit these repositories for the prediction and identification of proteotypic peptides. Characteristic physicochemical properties [22] of these peptides were largely taken into account to distinguish proteotypic peptides from other peptides. Around 500 physicochemical properties including charge, hydrophobicity, and secondary structure propensity used to discriminate proteotypic peptides from other peptides. Using this approach, >16,000 proteotypic peptides were identified for >4000 distinct yeast proteins. PeptideSieve is one such tool for the prediction of proteotypic peptides based mainly on the physicochemical propensity of the target peptides. PeptideSieve is available both as command line tool and as GUI windows version. The input can be either in the form of protein sequences in the FASTA file or TXT file of peptide sequences. The program first known to perform in silico digestion of proteins into peptides and based on its physicochemical properties computes likelihood function distinguishing proteins peptides to be proteotypic or not from rest of the peptides. The PeptideSieve tools can be downloaded and installed from the sashimi project at sourceforge (https://sourceforge.net/projects/sashimi/files/peptideSieve/). Further, these physicochemical properties were used in the development of algorithm for the prediction of proteotypic peptides using machine learning approaches [23, 24].

3 Discussion

The characterization of entire proteome, unlike genome, is very challenging due to dynamic nature of the proteome which rapidly changes due to change in the physiological state of the cells and its surrounding microenvironment. As a result, no complete characterization of proteome is available to date in spite of many recent advances in the mass spectrometry instruments which generates and identifies thousands of proteins from the complex mixture of protein samples. The cost and time involved in synthesizing thousands of candidates of clinical importance can be drastically reduced by specific set of peptides which uniquely represent target proteins. The proteomic repositories and computational proteomic tools have major role in identification of proteotypic peptides of target proteins. Further, these public proteomic repositories serve as important resource for creating proteotypic peptide libraries which can be used for the identification of proteins from tandem mass spectra to aid proteomics research community.

References

1. Whiteaker JR, Lin C, Kennedy J, Hou L, Trute M, Sokal I, Yan P, Schoenherr RM, Zhao L, Voytovich UJ, Kelly-Spratt KS, Krasnoselsky A, Gafken PR, Hogan JM, Jones LA, Wang P, Amon L, Chodosh LA, Nelson PS, McIntosh MW, Kemp CJ, Paulovich AG (2011) A targeted proteomics-based pipeline for verification of biomarkers in plasma. Nat Biotechnol 29(7):625–634. doi:10.1038/nbt.1900
2. Cohen Freue GV, Meredith A, Smith D, Bergman A, Sasaki M, Lam KK, Hollander Z, Opushneva N, Takhar M, Lin D, Wilson-McManus J, Balshaw R, Keown PA, Borchers CH, McManus B, Ng RT, McMaster WR, Biomarkers in T, the NCECPoOFCoET (2013) Computational biomarker pipeline from discovery to clinical implementation: plasma proteomic biomarkers for cardiac transplantation. PLoS Comput Biol 9(4), e1002963. doi:10.1371/journal.pcbi.1002963
3. Huttenhain R, Malmstrom J, Picotti P, Aebersold R (2009) Perspectives of targeted mass spectrometry for protein biomarker verification. Curr Opin Chem Biol 13(5-6):518–525. doi:10.1016/j.cbpa.2009.09.014
4. Brewis IA, Brennan P (2010) Proteomics technologies for the global identification and quantification of proteins. Adv Protein Chem Struct Biol 80:1–44. doi:10.1016/B978-0-12-381264-3.00001-1
5. Percy AJ, Chambers AG, Yang J, Hardie DB, Borchers CH (2014) Advances in multiplexed MRM-based protein biomarker quantitation toward clinical utility. Biochim Biophys Acta 1844(5):917–926. doi:10.1016/j.bbapap.2013.06.008
6. Lange V, Picotti P, Domon B, Aebersold R (2008) Selected reaction monitoring for quantitative proteomics: a tutorial. Mol Syst Biol 4:222. doi:10.1038/msb.2008.61
7. Picotti P, Aebersold R (2012) Selected reaction monitoring-based proteomics: workflows, potential, pitfalls and future directions. Nat Methods 9(6):555–566. doi:10.1038/nmeth.2015
8. Dittrich J, Becker S, Hecht M, Ceglarek U (2015) Sample preparation strategies for targeted proteomics via proteotypic peptides in human blood using liquid chromatography tandem mass spectrometry. Proteomics Clin Appl 9(1-2):5–16. doi:10.1002/prca.201400121
9. Vandemoortele G, Staes A, Gonnelli G, Samyn N, De Sutter D, Vandermarliere E, Timmerman E, Gevaert K, Martens L, Eyckerman S (2016) An extra dimension in protein tagging by quantifying universal proteotypic peptides using targeted proteomics. Sci Rep 6:27220. doi:10.1038/srep27220
10. Deutsch EW, Lam H, Aebersold R (2008) PeptideAtlas: a resource for target selection for emerging targeted proteomics workflows. EMBO Rep 9(5):429–434. doi:10.1038/embor.2008.56
11. Shi T, Song E, Nie S, Rodland KD, Liu T, Qian WJ, Smith RD (2016) Advances in targeted proteomics and applications to biomedical research. Proteomics. doi:10.1002/pmic.201500449
12. Song X, Amirkhani A, Wu JX, Pascovici D, Zaw T, Xavier D, Clarke SJ, Molloy MP (2016) Analytical performance of nanoLC-SRM using non-depleted human plasma over an 18-month period. Proteomics. doi:10.1002/pmic.201500507
13. Liebler DC, Zimmerman LJ (2013) Targeted quantitation of proteins by mass spectrometry. Biochemistry 52(22):3797–3806. doi:10.1021/bi400110b

14. Chambers AG, Percy AJ, Yang J, Borchers CH (2015) Multiple reaction monitoring enables precise quantification of 97 proteins in dried blood spots. Mol Cell Proteomics 14(11): 3094–3104. doi:10.1074/mcp.O115.049957
15. Craig R, Cortens JP, Beavis RC (2005) The use of proteotypic peptide libraries for protein identification. Rapid Commun Mass Spectrom 19(13):1844–1850. doi:10.1002/rcm.1992
16. Deutsch EW, Mendoza L, Shteynberg D, Slagel J, Sun Z, Moritz RL (2015) Trans-proteomic pipeline, a standardized data processing pipeline for large-scale reproducible proteomics informatics. Proteomics Clin Appl 9(7-8):745–754. doi:10.1002/prca.201400164
17. Farrah T, Deutsch EW, Hoopmann MR, Hallows JL, Sun Z, Huang CY, Moritz RL (2013) The state of the human proteome in 2012 as viewed through PeptideAtlas. J Proteome Res 12(1): 162–171. doi:10.1021/pr301012j
18. Vizcaino JA, Foster JM, Martens L (2010) Proteomics data repositories: providing a safe haven for your data and acting as a springboard for further research. J Proteomics 73(11):2136–2146. doi:10.1016/j.jprot.2010.06.008
19. Pan S, Aebersold R, Chen R, Rush J, Goodlett DR, McIntosh MW, Zhang J, Brentnall TA (2009) Mass spectrometry based targeted protein quantification: methods and applications. J Proteome Res 8(2):787–797. doi:10.1021/pr800538n
20. Farrah T, Deutsch EW, Kreisberg R, Sun Z, Campbell DS, Mendoza L, Kusebauch U, Brusniak MY, Huttenhain R, Schiess R, Selevsek N, Aebersold R, Moritz RL (2012) PASSEL: the PeptideAtlas SRMexperiment library. Proteomics 12(8):1170–1175. doi:10.1002/pmic.201100515
21. Shannon P, Markiel A, Ozier O, Baliga NS, Wang JT, Ramage D, Amin N, Schwikowski B, Ideker T (2003) Cytoscape: a software environment for integrated models of biomolecular interaction networks. Genome Res 13(11):2498–2504. doi:10.1101/gr.1239303
22. Mallick P, Schirle M, Chen SS, Flory MR, Lee H, Martin D, Ranish J, Raught B, Schmitt R, Werner T, Kuster B, Aebersold R (2007) Computational prediction of proteotypic peptides for quantitative proteomics. Nat Biotechnol 25(1):125–131. doi:10.1038/nbt1275
23. Webb-Robertson BJ, Cannon WR, Oehmen CS, Shah AR, Gurumoorthi V, Lipton MS, Waters KM (2010) A support vector machine model for the prediction of proteotypic peptides for accurate mass and time proteomics. Bioinformatics 26(13):1677–1683
24. Fusaro VA, Mani DR, Mesirov JP, Carr SA (2009) Prediction of high-responding peptides for targeted protein assays by mass spectrometry. Nat Biotechnol 27(2):190–198. doi:10.1038/nbt.1524

Chapter 9

Statistical Evaluation of Labeled Comparative Profiling Proteomics Experiments Using Permutation Test

Hien D. Nguyen, Geoffrey J. McLachlan, and Michelle M. Hill

Abstract

Comparative profiling proteomics experiments are important tools in biological research. In such experiments, tens to hundreds of thousands of peptides are measured simultaneously, with the goal of inferring protein abundance levels. Statistical evaluation of these datasets are required to determine proteins that are differentially abundant between the test samples. Previously we have reported the non-normal distribution of SILAC datasets, and demonstrated the permutation test to be a superior method for the statistical evaluation of non-normal peptide ratios. This chapter outlines the steps and the *R* scripts that can be used for performing permutation analysis with false discovery rate control via the Benjamini–Yekutieli method.

Key words Comparative profiling, Simultaneous testing, SILAC, Hypothesis test, Permutation test, False discovery rate

1 Introduction

The comparative profiling (shotgun) proteomics (CPP) experiment is now a staple research tool regularly employed to reveal differentially abundant proteins in biological samples. In such experiments, data-dependent acquisition on the mass spectrometer is performed for simultaneous protein identification and quantitation. Relative peptide quantitation is derived based on label free [1], or labeling approaches including SILAC (stable isotope labeling by amino acids in cell culture) [2], iTRAQ (isobaric tags for relative and absolute quantitation) [3], ICAT (isotope-code affinity tags) [4], and TMT (tandem mass tags) [5]. Although the method described herein is applicable to all the labeled techniques, we illustrate our methodology via the SILAC experimental setting.

In a typical two-plex SILAC experiment, cellular samples were labeled with "light" and "heavy" isotopes via feeding the cells or organisms with amino acids that contain the required isotopes. Lysates or subcellular fractions are then prepared and combined, usually based on equal protein content. The target cellular proteome

Shivakumar Keerthikumar and Suresh Mathivanan (eds.), *Proteome Bioinformatics*, Methods in Molecular Biology, vol. 1549, DOI 10.1007/978-1-4939-6740-7_9, © Springer Science+Business Media LLC 2017

is isolated, digested with trypsin, and the peptides are then analyzed via tandem mass spectrometry (MS/MS). Due to the predictability of the mass shift between the peptides that are labeled as "light" or "heavy," it is possible to distinguish such labels from survey mass spectra. This allows for the calculation of relative intensity ratios. Here, a peptide ratio equal to one would indicate no difference between the labeled populations, while a ratio that is different from one would indicate an upregulation or downregulation of the quantified peptide.

Using fragments of parent ions (from the MS/MS spectra), it is also possible match the quantified peptides to protein sequences, and conduct inference regarding protein abundances via averaging over the matched peptides. Peptide-to-protein matching is generally performed by specialist software; *see* Lau et al. [6] and Rigbolt and Blagoev [7] for details. We refer to Mann [8] for a comprehensive treatment of the applications of SILAC experiments. In this chapter, we are concerned with the statistical assessment of the relative abundance quantitation data.

CPP experiments often infer the abundance ratios of thousands of proteins from tens of thousands of peptide ratios. It is important to process these ratios in a coherent and statistically valid manner. Quantitative experiments including SILAC often rely on fold-change cutoffs for selection of significantly altered proteins. The selection of fold-change cut-offs is often arbitrary, experiment-dependent, and do not account for the variability or distribution of the ratios. For example, depending on the experiment, thresholds between 1.3 [8] and 6 [2] have been suggested in the literature. A more nuanced and statistically valid approach is to assess the significance of peptide and protein ratios via hypothesis tests, such as t-tests, z-tests, and Wilcoxon signed-rank tests. There are several potential pitfalls in applying such methodologies to the assessment of peptide ratios, including incorrect assumptions and deficiencies in power [9]. In comparison, a permutation test-based approach, which does not make assumptions on the normality or independence of the observed data was benchmarked against the existing methodology, and found to be more powerful and robust [9]. To facilitate application of the permutation test in quantitative proteomics experiments, we implemented a free web-based tool which allows user-selected parameters [10]. In this chapter, we provide instructions and the R script to enable researchers the flexibility to customize the analysis.

Aside from the assessment of significance of individual protein ratios, the large-scale and simultaneous nature of CPP necessitates the mitigation against false-positive results. Here, we provide an implementation FDR control algorithm of Benjamini and Yekutieli [11], via an R script. The FDR control methodology is the same as that which is used in the online QPPC tool of Chen et al. [10].

2 Materials

1. Software tools: A spreadsheet software such as *Excel* (*Microsoft*) is required to manipulate the raw data. The *R* statistical computing environment is used to perform all statistical computations. It is available free at https://cran.r-project.org or https://www.rstudio.com (*see* **Note 1**).
2. Data: The starting data for a permutation-test based quantification of data resulting from a CPP experiment is a peptide summary report that records a peptide ratio in each row of the data, as well as the protein accession to which that peptide assigned. Such reports may be generated from various software packages for mass spectrometry data analysis, for example, 'peptide summary export' from *Spectrum Mill.* All biological replicates should be combined into a single file, either during the export, or within Excel (*see* **Note 2**).

3 Methods

In this Chapter, a publically available data set, from Chen et al. [10] will be used to illustrate the permutation test using R script (Fig. 1). This sample dataset can be downloaded from http://qppc.di.uq.edu.au/docs/Sample_dataset.csv.

Prior to permutation testing, the data file must be converted to the required format with specific headings as exemplified in Fig. 2. The file should be in the .csv (comma-separated values) format. The raw data is required to have at least two columns; the first

	A	B	C
1	accession_number	entry_name	L/H
2	P21589	5'-nucleotidase	0.784
3	P62873	Guanine nucleotide-binding protein G(I)/G(S)/G(T) subunit beta-1	1.333
4	P21589	5'-nucleotidase	0.908
5	Q71U36	Tubulin alpha-1A chain	2.802
6	P21589	5'-nucleotidase	1.306
7	P62873	Guanine nucleotide-binding protein G(I)/G(S)/G(T) subunit beta-1	1.396
8	P60709	Actin, cytoplasmic 1	1.575
9	P62873	Guanine nucleotide-binding protein G(I)/G(S)/G(T) subunit beta-1	1.33
10	Q9P121	Neurotrimin	0.418
11	P00387	NADH-cytochrome b5 reductase 3	0.723
12	Q03135	Caveolin-1	1
13	P21589	5'-nucleotidase	1.254
14	P21589	5'-nucleotidase	1.168
15	Q14254	Flotillin-2	2.349
16	P08754	Guanine nucleotide-binding protein G(k) subunit alpha	0.906
17	P62736	Actin, aortic smooth muscle	0.926
18	Q03135	Caveolin-1	1.407
19	Q71U36	Tubulin alpha-1A chain	1.776
20	P21589	5'-nucleotidase	1.126

Fig. 1 Screenshot of an Excel spreadsheet containing the column headers of the sample raw data file that is obtained from http://qppc.di.uq.edu.au/docs/Sample_dataset.csv

	A	B	C
1	protein	entry_name	ratio
2	P62258	14-3-3 protein epsilon	2.681
3	P62258	14-3-3 protein epsilon	1.299
4	P63104	14-3-3 protein zeta/delta	1.721
5	Q13200	26S proteasome non-ATPase regulatory subunit 2	1.175
6	Q13200	26S proteasome non-ATPase regulatory subunit 2	1.177
7	Q9UJ83	2-hydroxyacyl-CoA lyase 1	1.285
8	Q9UJ83	2-hydroxyacyl-CoA lyase 1	1.758
9	Q9UJ83	2-hydroxyacyl-CoA lyase 1	1.243
10	Q9UJ83	2-hydroxyacyl-CoA lyase 1	1.295
11	P15880	40S ribosomal protein S2	1.527
12	P08195	4F2 cell-surface antigen heavy chain	2.133
13	P21589	5'-nucleotidase	0.784
14	P21589	5'-nucleotidase	0.908
15	P21589	5'-nucleotidase	1.306
16	P21589	5'-nucleotidase	1.254
17	P21589	5'-nucleotidase	1.168
18	P21589	5'-nucleotidase	1.126
19	P21589	5'-nucleotidase	0.846
20	P21589	5'-nucleotidase	0.673

Fig. 2 Processed sample data showing the required column headers. The data is sorted by entry_name to enable quality inspection

column should contain an 'accession number'. This column indicates the protein to which the peptide on each row is matched. The second column required column is a numerical value that indicates the peptide ratio. The script will retain additional columns, for example, the same dataset contains an 'entry name' column (Figs. 1 and 2).

The main component of this chapter is an R script that conducts the permutation test and performs FDR control. The R script performs the following steps:

1. Get the unique proteins identified in the dataset.
2. Exclude the single observation proteins.
3. Compute the average log-ratio for each unique protein.
4. Perform N permutations for each unique protein.
5. Compute permutation p-values for each unique protein.
6. Control FDR at an appropriate, user-specified level.
7. Write a .csv file that contains the p-values.

3.1 Data Preprocessing

1. Open the raw data file in Excel.
2. Rename the 'accession number' column of the raw data to protein, the ratio column of interest to ratio (*see* **Note 3**).
3. Manual validation of the data quality should be performed, including the removal of rows if there are any negative, zero, or character strings (*see* **Note 4**). Known contaminant proteins based on biological knowledge of the experiment can also be removed, for example, albumin (*see* **Note 5**).
4. Save the document in .csv format, in a directory, which will be the working directory for R.

3.2 Permutation Test and FDR Control Using R

1. If required, install R Studio (*see* **Note 1**).
2. Set the working directory to the desired folder that contains the files and data that are required for the analysis. This can be done by selecting Session -> Set Working Directory -> Choose Directory ..., and select the desired directory.
3. Open a new script by selecting File -> New File -> R Script, or using the +document icon, or Ctrl+Shift+N on a Windows system.
4. Copy the script below to the file. Due to the difference in coding for quotation marks between *Word* and *R*, it is necessary to retype the ‘ ’ and “ “ symbols within *R* after copy-pasting. Also refer to **Notes 6–10**.

```
# -------------------------------------------------------
# R script for permutation test
# -------------------------------------------------------
# Inputs:
# Number of permutation samples.
N = 1000
# False Discovery Control Rate.
control = 0.10
# Location of CSV file that contains the raw data.
# CSV file must contain at least protein and ratio cols.
loc = "https://www.dropbox.com/s/7i2b1nu618dzu7s/MIMB_SM_NMH_2015.
csv?dl=1"
# Read the CSV file and store as a data frame.
data = read.csv(loc)
# Get the unique proteins that are identified in the data.
unique.proteins = unique(data$protein)
# Remove proteins with only one observation
no.one.proteins = as.factor(c())
for (j in 1:length(unique.proteins)) {
 if (length(which(data$protein==unique.proteins[j]))>1) {
   no.one.proteins = c(no.one.proteins,j)
   }
 }
 unique.proteins = unique.proteins[no.one.proteins]
 # Compute the average log-ratio for each unique protein.
 log.ratios = c()
 data$ratio = as.numeric(as.character(data$ratio))
 for (j in 1:length(unique.proteins)) {
   log.ratios[j] = mean(log(data$ratio[
    which(data$protein==unique.proteins[j])]))
 }
 # Compute permutation p-value for each unique protein.
 p.values = c()
 for (j in 1:length(unique.proteins)) {
  ## Get number of peptides that are matched to protein j.
  number = length(which(data$protein==unique.proteins[j]))
```

```
  ## Make a variable to store r_tilde_k values.
  r.tilde = c()
  ## Perform permutations.
  for (k in 1:N) {
   ## Obtain a random permutation sample.
   perm.sample = sample(data$ratio,number,replace=T)
   ## Compute the permutation sample average log-ratio.
   r.tilde[k] = mean(log(perm.sample))
 }
 ## Compute p-value from the permutation samples.
 p.values[j] = (sum(abs(r.tilde)>abs(log.ratios[j]))+1)/
  (N+1)
}
# Compute averate ratios.
av.ratios = exp(log.ratios)
# Perform FDR control at the specified 'control' level.
fdr = as.numeric(p.adjust(p.values,method='BY') < control)
# Write a CSV file that contains the p-values.
csv.file = as.data.frame(cbind(as.character(
 unique.proteins),log.ratios,av.ratios,p.values,fdr))
colnames(csv.file) = c('protein','av.log.ratio','av.ratio',
                 'p.value','fdr.significant')
write.csv(csv.file,file='output.csv',row.names=F)
```

1. Save the script as "permutation_test.R".
2. The script contains three user inputs: the number of permutations *N*, the location of the .csv preprocessed raw data file loc, and the FDR control level. **Note 7** provides advice regarding the setting of *N*, for the setting of loc and control (*see* **Notes 6** and **8**, respectively).
3. In the Console window, type the command source ('permutation_test.R') to execute the permutation test script. Note that internet connection is necessary to retrieve the online document as specified in the script.
4. The output results of the R script are in the .csv file output.csv. *See* **Note 9** regarding the interpretation of columns.
5. For reference, the first ten entries of the output file from the sample data is shown in Table 1.

4 Notes

1. The *R* statistical environment is available for all three major operating systems (i.e., Mac, Linux, and Windows). Although comprehensive and powerful, the standard *R* environment is bare and may be an unappealing work environment for new users. We recommend the *RStudio* environment for a more user-friendly

Table 1
Permutation test results for the first ten proteins (in alphabetical order by UniProt accession), from the preprocessed sample data that is available from http://tinyurl.com/hiendnguyen-MIMB-SM-NMH-2015

Protein	av.log.ratio	av.ratio	*p*-Value	fdr.significant
O75477	−0.198433735	0.820014107	0.450549451	0
O75695	0.493099779	1.637383889	0.12987013	0
O75955	0.613125124	1.846191971	0.000999001	1
O94905	0.252036067	1.286642441	0.290709291	0
P00387	−0.11016517	0.895686183	0.7002997	0
P04156	−0.091557763	0.912508606	0.769230769	0
P04216	−0.187514028	0.829017489	0.427572428	0
P04406	0.528803331	1.696900465	0.071928072	0
P04899	0.058877585	1.060645394	0.89010989	0

experience. The *RStudio* software is available for all three major operating systems and can be downloaded from https://www.rstudio.com.

2. In general, CPP experiments employ biological replication rather than technical replication. It is not recommended to combine biological and technical replicates.
3. In the outputs of a typical two-plex SILAC experiment, there is only one ratio column, either labeled L/H or H/L. In a multiplex (e.g., three-plex) experiment, one is required to identify which of the ratios is of current interest. For example, it is common to have the ratios labeled as L/H, L/M, and M/H (here M stands for medium). For each permutation analysis, select the ratio of interest, out of the three, and change it to ratio.
4. In *Excel*, a simple method for exclusion of non-positive-numeric values is to use the sort tool on the ratio column of the raw data. If the data is sorted 'A to Z', then the rows that contain negative or zero ratios should appear at the top of the table. Rows that contain character-string values should appear at the bottom of the table. Both of these sets of rows can be deleted from the raw data table, before further processing.
5. A small number of contaminants is unlikely to impact the estimation of permutation ratios. However, if a large number of keratins are evident, it would be prudent to evaluate p-values with and without removal of the potential contaminants.

6. When reading the script, it is important to note that the symbol # denotes comments, in the R language. These lines are not parse as executable script but rather informs the programmer/ reader of the intent of the script.

   The script is written to take as an input the sample preprocessed data file that is available from http://tinyurl.com/hiendnguyen-MIMB-SM-NMH-2015. To input a file (e.g., XXX.csv) that is stored in the desired working directory (*see* Subheading 3.2), edit the script to read loc = "XXX.csv" where loc appears for the first time, in the script. As an alternative, to the use of the loc = "XXX.csv" and unique.proteins = unique (data$protein), within the script, one can use the 'Import Dataset' command in R Studio. To import a dataset to the working environment using the 'Import Dataset' command, click the 'Import Dataset Icon' under the 'Environment' tab and navigating to the correct file (Fig. 3a). Import the dataset as "data" using default parameters. The dataset can be visualized within R (Fig. 3b).

7. A recommended starting value of $N = 1000$ was empirically determined, which allows for sufficient accuracy of the *p*-value estimates and does not utilize excessive computation resources [10].

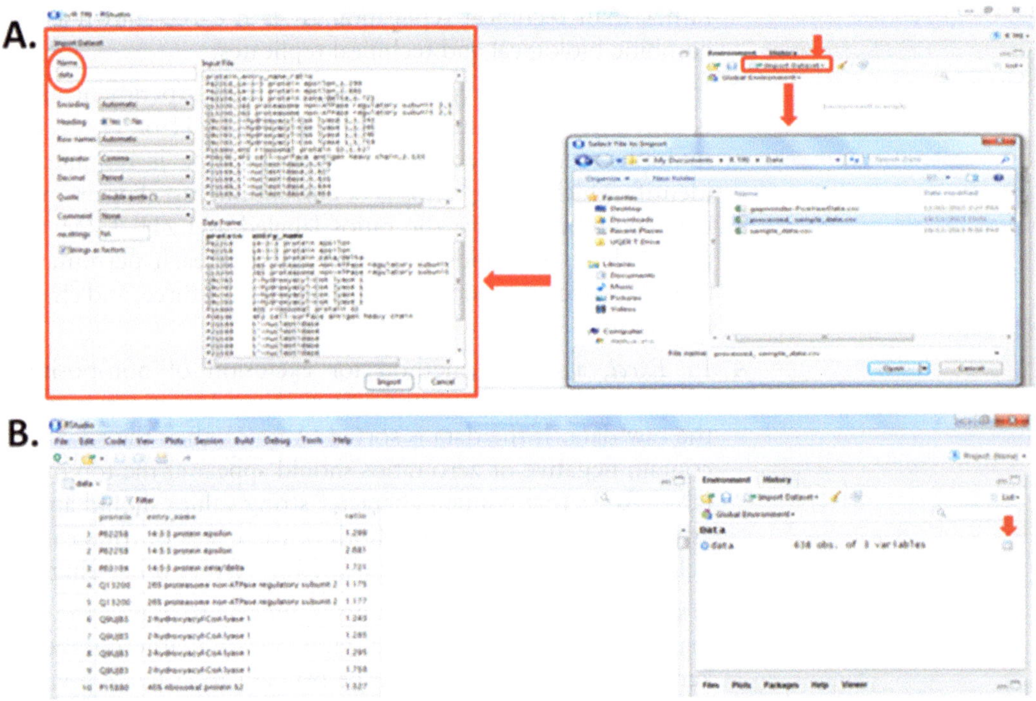

Fig. 3 Importing processed data in R Studio. (**a**) Use the Import Dataset tool (*arrow*) and selecting the processed data file using the explorer window shown on the *lower right*. (**b**) View the imported data (*left window*) using the spreadsheet icon (*arrow*)

8. It is conventional to control the FDR at a level that is comparable to the significance level of a single hypothesis test. We set the default level at 10%, but it is also common to observe FDR controlled at the 5% level in many experiments. To change the level at which FDR is controlled (e.g., 5%), edit the script to read control = 0.05 where control appears for the first time, in the script.
9. The output is a .csv file that contains three columns: protein, av.log.ratio, av.ratio, p.value, and fdr.significant. These columns contain the name of the unique proteins from the experiment, and the average log-ratio, average ratio, *p*-value for that protein, and whether the protein is significant at the FDR control level, respectively. The average ratio for each protein is computed as the exponential (i.e., *exp* in *Excel* and *R*) of the av. log-ratio column. The column fdr.significant takes values of 0 or 1; 1 indicates that the protein is significant at the control level, and 0 indicates otherwise.
10. The R script provided makes the implicit assumption that proteins observed via a single peptide are unreliable observations. The script will only conduct the permutation test analysis on proteins observed via two or more peptides. In order to quantitate proteins that are singly observed, delete lines 16–23.

References

1. Bantscheff M, Schirle M, Sweetman G, Rick J, Kuster B (2007) Quantitative mass spectrometry in proteomics: a critical review. Anal Bioanal Chem 389(4):1017–1031. doi:10.1007/s00216-007-1486-6
2. Ong SE, Blagoev B, Kratchmarova I, Kristensen DB, Steen H, Pandey A, Mann M (2002) Stable isotope labeling by amino acids in cell culture, SILAC, as a simple and accurate approach to expression proteomics. Mol Cell Proteomics 1(5):376–386
3. Ross PL, Huang YN, Marchese JN, Williamson B, Parker K, Hattan S, Khainovski N, Pillai S, Dey S, Daniels S, Purkayastha S, Juhasz P, Martin S, Bartlet-Jones M, He F, Jacobson A, Pappin DJ (2004) Multiplexed protein quantitation in Saccharomyces cerevisiae using amine-reactive isobaric tagging reagents. Mol Cell Proteomics 3(12):1154–1169. doi:10.1074/mcp.M400129-MCP200
4. Gygi SP, Rist B, Gerber SA, Turecek F, Gelb MH, Aebersold R (1999) Quantitative analysis of complex protein mixtures using isotope-coded affinity tags. Nat Biotechnol 17(10):994–999. doi:10.1038/13690
5. Thompson A, Schafer J, Kuhn K, Kienle S, Schwarz J, Schmidt G, Neumann T, Johnstone R, Mohammed AK, Hamon C (2003) Tandem mass tags: a novel quantification strategy for comparative analysis of complex protein mixtures by MS/MS. Anal Chem 75(8):1895–1904
6. Lau KW, Jones AR, Swainston N, Siepen JA, Hubbard SJ (2007) Capture and analysis of quantitative proteomic data. Proteomics 7(16):2787–2799. doi:10.1002/pmic.200700127
7. Rigbolt KT, Blagoev B (2010) Proteome-wide quantitation by SILAC. Methods Mol Biol658:187–204.doi:10.1007/978-1-60761-780-8_11
8. Mann M (2006) Functional and quantitative proteomics using SILAC. Nat Rev Mol Cell Biol 7(12):952–958. doi:10.1038/nrm2067
9. Nguyen H, Wood IA, Hill MM (2012) A robust permutation test for quantitative SILAC proteomics experiments. J Integr Omics 2:80–93
10. Chen D, Shah A, Nguyen H, Loo D, Inder KL, Hill MM (2014) Online quantitative proteomics p-value calculator for permutation-based statistical testing of peptide ratios. J Proteome Res 13(9):4184–4191. doi:10.1021/pr500525e
11. Benjamini Y, Yekutieli D (2001) The control of the false discovery rate in multiple testing under dependency. Ann Stat 29:1165–1188

Chapter 10

De Novo Peptide Sequencing: Deep Mining of High-Resolution Mass Spectrometry Data

Mohammad Tawhidul Islam, Abidali Mohamedali, Criselda Santan Fernandes, Mark S. Baker, and Shoba Ranganathan

Abstract

High resolution mass spectrometry has revolutionized proteomics over the past decade, resulting in tremendous amounts of data in the form of mass spectra, being generated in a relatively short span of time. The mining of this spectral data for analysis and interpretation though has lagged behind such that potentially valuable data is being overlooked because it does not fit into the mold of traditional database searching methodologies. Although the analysis of spectra by de novo sequences removes such biases and has been available for a long period of time, its uptake has been slow or almost nonexistent within the scientific community. In this chapter, we propose a methodology to integrate de novo peptide sequencing using three commonly available software solutions in tandem, complemented by homology searching, and manual validation of spectra. This simplified method would allow greater use of de novo sequencing approaches and potentially greatly increase proteome coverage leading to the unearthing of valuable insights into protein biology, especially of organisms whose genomes have been recently sequenced or are poorly annotated.

Key words *De novo* peptide sequencing, Hybrid peptide sequencing, MS validation, MS evidence, Functional annotation

1 Introduction

Proteomics, as a science, has progressed in leaps and bounds in the past decade on the back of powerful and sophisticated mass spectrometry technologies and computing power. The fundamental principal in protein identification from mass spectrometry data is the peptide mass fingerprint (PMF) where spectra are matched to an existing database of theoretical spectra based on sequence data from genomic studies. This data is then scored, based on confidence to reveal the most reliable matches [1].

The fundamental requirement of this approach is that a reliable sequence database in a reasonably mature form (with open reading frames or ORFs, splice variants, mutations, etc.) must exist for matching to occur. If this does not occur, not only does this

Shivakumar Keerthikumar and Suresh Mathivanan (eds.), *Proteome Bioinformatics*, Methods in Molecular Biology, vol. 1549, DOI 10.1007/978-1-4939-6740-7_10, © Springer Science+Business Media LLC 2017

hamper the investigation of the proteomes of non-model organisms whose genomes have yet to be sequenced, it may also mean that ambiguities in genomic sequences, such as alternative splice sites, mutations [2], and ambiguous and alternative ORFs may curtail the investigation of known genomes. Indeed, one of the very reasons that over 2900 proteins are currently considered "missing" in the human proteome could be that their spectra simply could not match the sequences found in any of the known databases. One of the ways in which this fundamental problem can be addressed is the use of de novo sequencing, where the peptide sequence (with or without PTMs) can be elucidated independently from the distance between peaks in a mass spectrum [3].

Although the earliest reports of de novo peptide sequencing predates database searching [4], with the earliest computer-aided programs described in the 1970s, it is only more recently, with significant advances in computation and in high resolution mass spectrometry, de novo sequencing algorithms are being used more widely. The most common criticisms of de novo sequencing are that the data produced is not as reliable as that from database searching and that no reliable FDR exists to discern sequenced data. Several approaches have been used to address this issue, such as the use of isotopic labeling of peptides [3], combining different fragmentation techniques [5], more innovative sequencing algorithms [6], combining de novo sequencing workflows with homology searches [7], and many more. Some of these methodologies increase confidence with most requiring some form of modification of experimental criteria, making working on archival results difficult.

To date, numerous computer programs using different algorithms for sequencing have been published some of which are freely available such as Lutefisk [8], PepNovo [9], PEAKS [10], NovoHMM [11] and UniNovo [12]. We describe here a methodology to investigate mass spectrometry data using three common de novo sequencing software, many of which are freely accessible to the academic community. We demonstrate that in tandem with database searching, and homology matching coupled with the utilization of at least three algorithms, de novo peptide sequencing can greatly increase the overall coverage of the proteome in a reliable and accurate manner (Fig. 1). We propose further that de novo sequencing cannot be a standalone solution, as accurate manual spectral validation (as presented elsewhere in this book [13]) is a prerequisite for increased confidence in proteomics data.

2 Materials

2.1 Data Sources

2.1.1 De Novo Peptide Sequencing

1. Raw nanoLC-ESI-MS/MS data files in .mgf format (*see* **Note 1**).
2. List of proteins identified by Mascot search [1] using the same raw data (i.e., from Subheading 2.1.1).

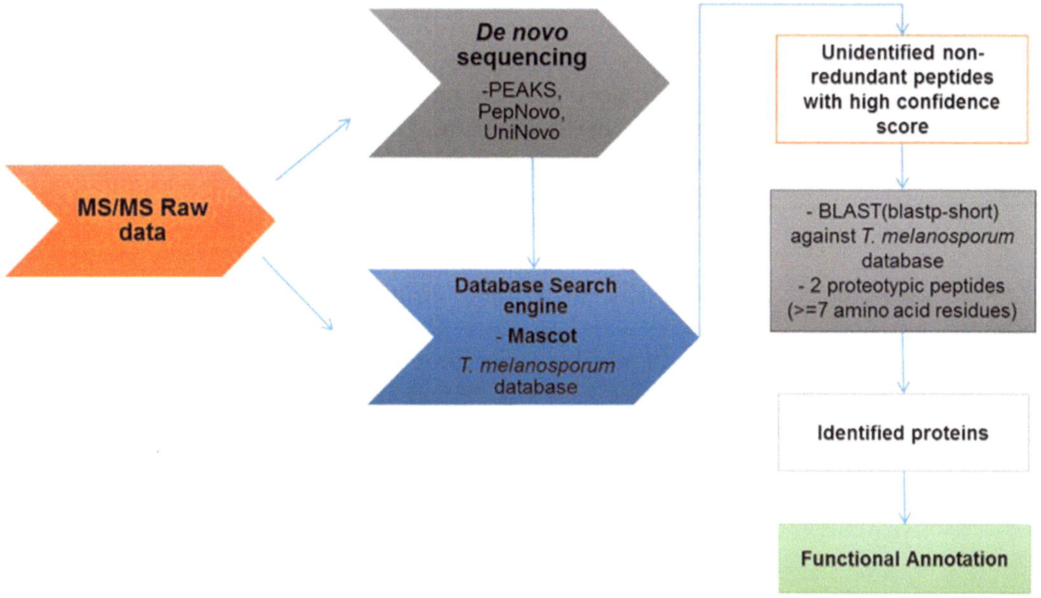

Fig. 1 Outline of the overall workflow

3. Nonredundant Black Perigord truffle (*Tuber melanosporum* Vittad.) protein sequences from the MycorWeb database [14, 15].

2.1.2 Sequence Similarity and Functional Annotation

1. Download the following protein datasets in FASTA format from UniProtKB/Swiss-Prot database [16].
   (a) Yeast proteins with experimental evidence.
   (b) Fungal proteins with experimental evidence.
   (c) Reviewed fungal proteins .
2. Protein Data Bank (PDB) [17] proteins from the PDB.

2.2 Software

1. De novo peptide sequencing tools:
   (a) PEAKS[10].
   (b) PepNovo [9].
   (c) UniNovo [12].
2. BLAST [18] package of tools for database similarity search to identify proteins homologous to predicted protein sequences, known proteins or structures as well as for creating specialized search databases from sequence datasets [19].
3. InterProScan [20] for protein domains mapping [21].
4. KAAS[22] or KOBAS 2.0 [23] for pathway mapping.

3 Methods

3.1 De Novo Sequencing of Peptides

Peptide de novo sequencing derives an amino acid sequence from its tandem mass spectrum (MS/MS) based on the distances between peaks using algorithms to account for charge states, modifications, cleavage sites, probabilistic assessments and other factors [24, 25]. This method is outlined in Fig. 2.

3.1.1 De Novo Peptide Sequencing with PEAKS

PEAKS uses a dynamic programming algorithm to compute sequences with best score. It computes the best possible sequence among all possible amino acid combinations. It connects each output sequence with a score and also provides positional confidence scores to determine the correct sequences or amino acids [10].

1. First step of the *de novo* sequencing is to create a project. To create a project, open PEAKS Studio.
2. Then click File →**New Project** which will start the project wizard shown in Fig. 3.

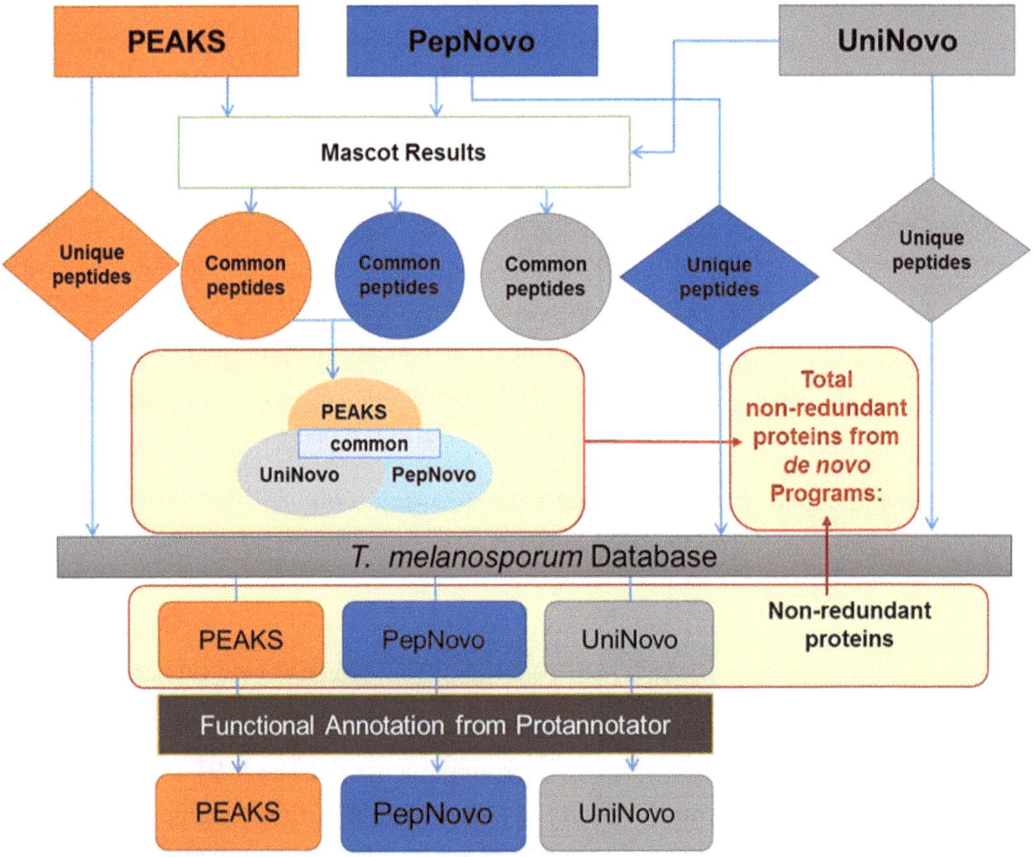

Fig. 2 Workflow illustrating the methodology for protein identification using de novo peptide sequencing and functional annotation

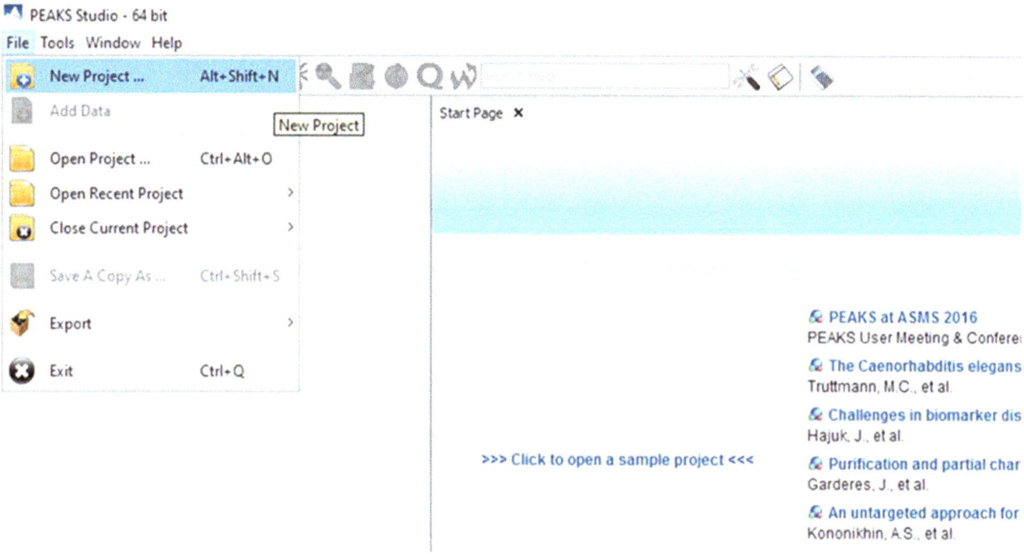

Fig. 3 Opening a new project in PEAKS 7.5

3. Enter the project name (1), select a location (2) for the project, browse data files (7) to add data files to the project (4), and then click **Data Refinement** (5), as shown in Fig. 4.
4. Select the default options and then click **Finish**, as illustrated in Fig. 5. The data processing can take some time to run (*see* **Note 2**).
5. Once the data processing is completed, go to the projects listed and Right click on your project, followed by clicking on *De novo*, as shown in Fig. 6.
6. Select **trypsin** as the enzyme, **carbamidomethylation** as fixed modification and **oxidation of methionine and deamination of asparagine and glutamine** as variable modifications (alternative modifications may be selected dependent on how the sample was processed/unique biochemistry being investigated). Set the **precursor mass tolerance** to 0.05 Da, **fragment ion tolerance** to 0.03 Da (the sample data was run on an AbSciex 5600 Triple TOF unit, these parameters can be adjusted to suit other instruments/thresholds) and select **3 candidates/spectrum** with **1% false discovery rate** (FDR), as shown in Fig. 7. Then click **Finish** to run (*see* **Note 2**).
7. Once completed, you should see the *De novo* results at the end of your sample on the left hand site. *Double click* the result note to view the sequencing results.
8. Click the De novo Tab (as indicated in Fig. 8) to view the identified peptides along with their score, retention time, spectra, spectral annotation, and ion match table (*see* **Note 3**).

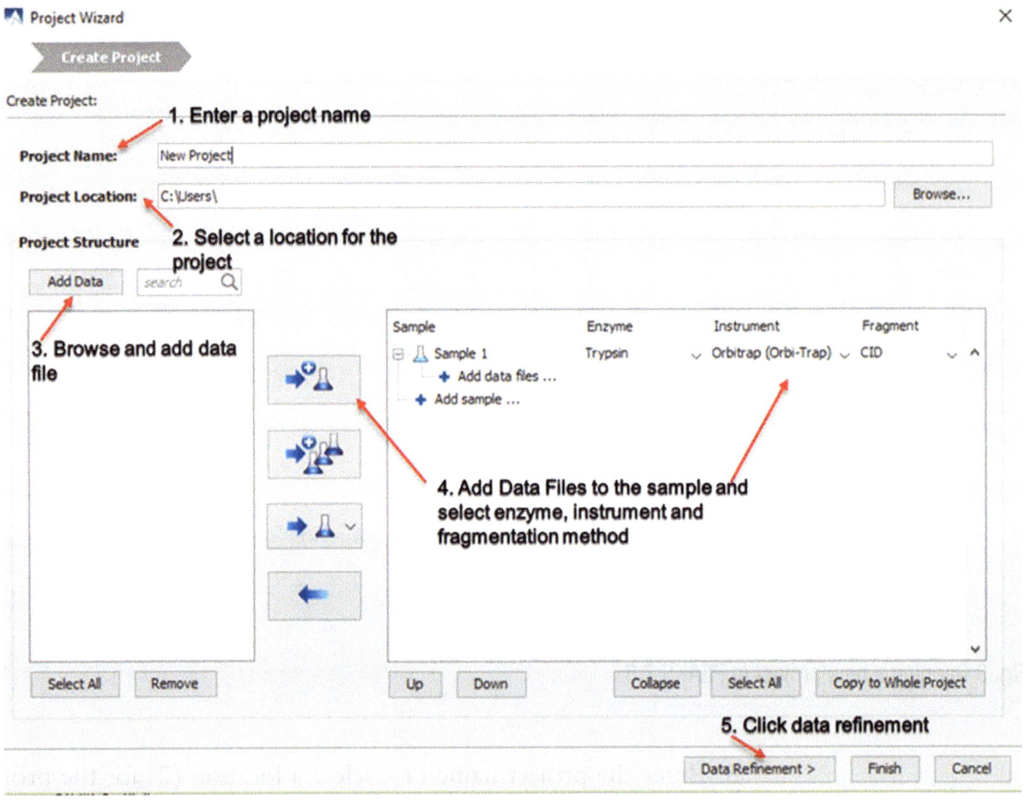

Fig. 4 Adding data obtained from a mass spectrometer for analysis

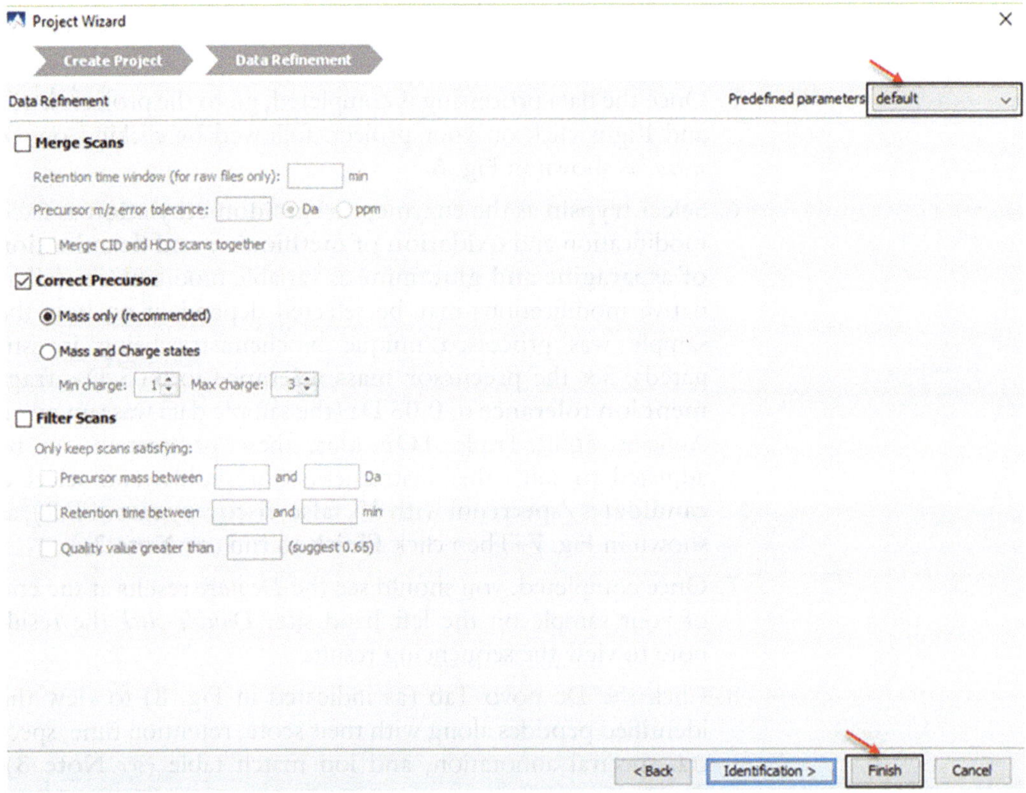

Fig. 5 Selecting **Data Refinement** options

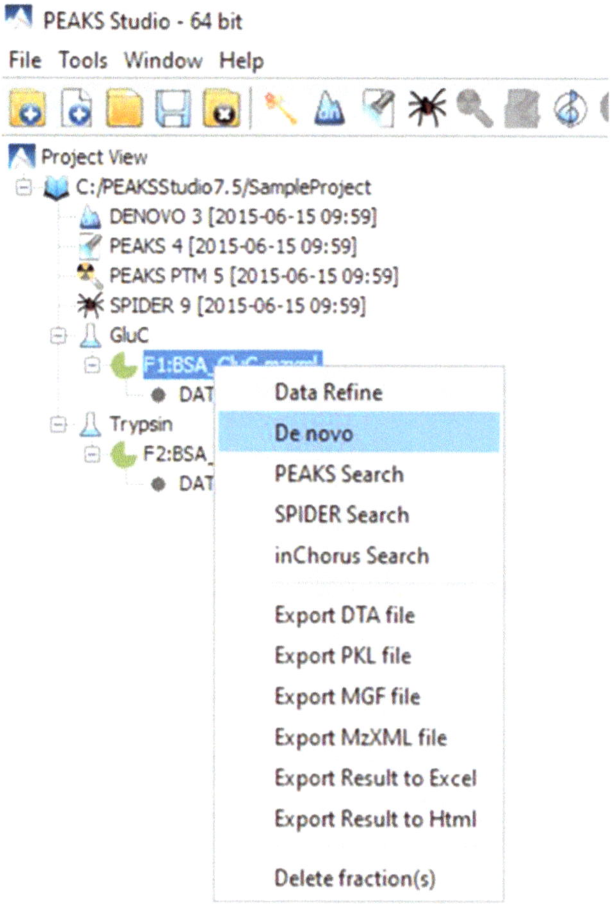

Fig. 6 Running the De novo function

Instructions on how to manually validate the spectra are provided elsewhere [13].

9. Go to the Summary tab, then set the Average Local Confidence value and export all peptides to a file as illustrated in Fig. 9 (*see* **Note 3**).

3.1.2 De Novo Peptide Sequencing with PepNovo

PepNovo is an ion-trap mass spectrometry data-specific de novo sequencing tool that works on spectral graph construction method, to determine the best possible score in a graph. It uses a probabilistic network to model peptide fragmentation events of mass spectrometers [9].

1. Download and install the latest PepNovo [9] tool on your machine.
2. Perform de novo peptide sequencing using the following parameters (*see* **Note 4**).

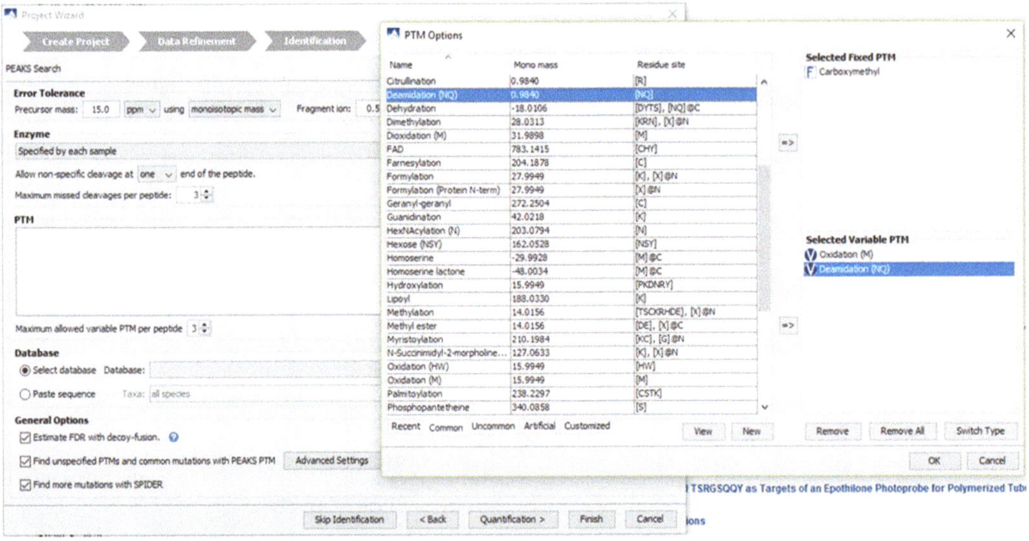

Fig. 7 Selecting fixed and variable modification options

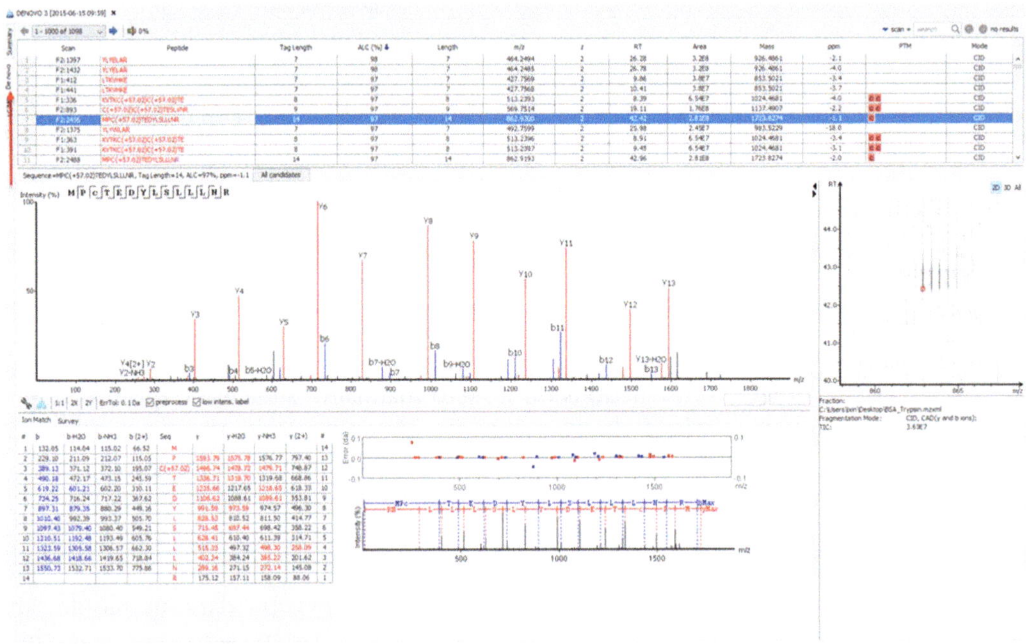

Fig. 8 De novo results for viewing spectra and associated information for manual spectral validation

Model (-model)= CID_IT_TRYP

PTMs (-PTMs)= C + 57:M + 16:N + 1

Fragment tolerance (-fragment_tolerance) = 0.03

Precursor mass tolerance (-pm_tolerance) = 0.05

Enzyme (-digest) = TRYPSIN

Sample command:

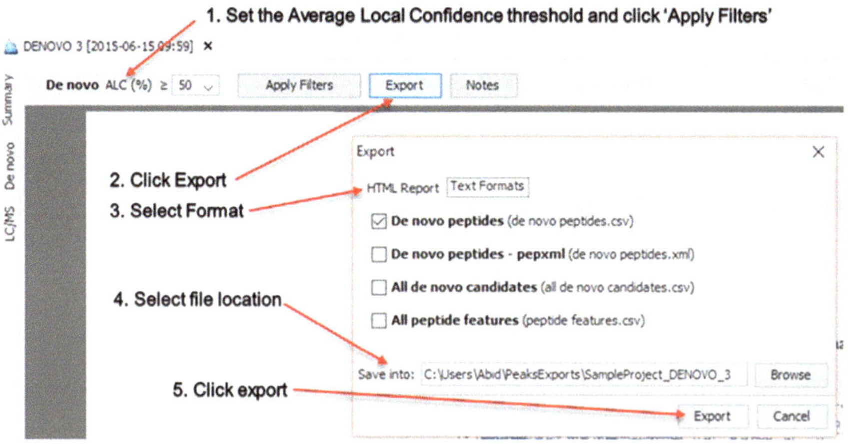

Fig. 9 Exporting data for further analysis

```
./PepNovo_bin -file </full/path/to/file/filename.
mgf> -model CID_IT_TRYP -PTMs C+57:M+16:N+1
-fragment_tolerance 0.03 -pm_tolerance 0.05
-digest TRYPSIN > <path/to/output/file>
```

Once completed, the output file should look like the image shown below:

```
PepNovo+ Build 20101117
Copyright 2010, The Regents of the University of California. All Rights Reserved.
Created by Ari Frank (arf@cs.ucsd.edu)

Initializing models (this might take a few seconds)... Done.
Fragment tolerance : 0.0300
PM tolernace       : 0.0500
PTMs considered    : C+57:M+16:N+1

>> 0 0 Locus:1.1.1.987.2 File:"Abid_24Aug2012_TP01.wiff" (SQS 0.5055)
#Index  RnkScr  PnvScr  N-Gap   C-Gap   [M+H]   Charge  Sequence
0       6.185   -41.564 115.061 0.000   849.512 2       AFVGANR
1       6.107   -41.049 115.061 0.000   849.512 2       FAVGANR
2       5.749   -41.726 115.061 0.000   849.512 2       FAVAGNR
3       5.590   -42.242 115.061 0.000   849.512 2       AFVAGNR
4       4.387   5.771   0.000   0.000   850.515 2       RLLGVHR
5       3.907   -19.825 115.061 0.000   850.515 2       GGSLFVR
6       3.774   -33.902 115.061 0.000   850.515 2       GGSLM+16VR
7       3.705   -19.993 0.000   0.000   850.515 2       RLM+16MRK
8       3.685   -20.095 0.000   0.000   850.515 2       RLM+16RMK
9       3.499   -20.598 115.061 0.000   850.515 2       GGSLVFR
10      3.498   4.573   0.000   0.000   850.515 2       RLLGHVR
11      3.448   -31.211 115.061 0.000   850.515 2       GGSLVM+16R
12      3.323   -41.122 128.132 0.000   851.518 2       AAGVSYR
13      3.226   -41.600 128.132 0.000   851.518 2       AAGVYSR
14      3.177   -27.643 115.061 0.000   850.515 2       GGTVFVR
15      3.151   0.726   0.000   0.000   850.515 2       RLVGFMK
16      3.088   -21.753 115.061 0.000   850.515 2       GGLSFVR
17      3.080   -2.731  0.000   0.000   850.515 2       RLVGMFK
18      3.003   -11.359 0.000   0.000   850.515 2       RLVGN+1YK
19      2.962   -16.514 0.000   0.000   850.515 2       QTLSFVR
```

3.1.3 De Novo Peptide Sequencing with UniNovo

UniNovo [12] is suitable for all types of mass spectral data, such as collision-induced dissociation (CID), higher-energy C-trap dissociation (HCD) and electron-transfer dissociation (ETD) spectra of trypsin, LysC or AspN digested peptides.

1. Download and install the latest UniNovo [12] tool on your machine.
2. Perform a de novo peptide sequencing using the following parameters (*see* **Notes 5** and **6**).

   Minimum length of reconstructions (−l) = 10

   Accuracy threshold (−acc) = 0.6

   Ion tolerances (6) = 0.03 Da

   Precursor ion tolerance (−pt) = 0.05Da

   Fragmentation methods (−f) = CID

   Enzyme applied (−e) = trypsin [1]

   Sample command:

```
java -jar UniNovo.jar -i <input file > -o <output file prefix> -l
10 -acc 0.6 -t 0.03Da -pt 0.05Da -f CID -e 1
```

Once the process is completed, you should see a screen like the one below:

```
>> 929  911.4674    2    null   0.7227796
>> 930  607.9841    0    null   0.0      filtered out 0
>> 931  607.9785    0    null   0.0      filtered out 0
>> 932  607.9826    3    null   0.0      filtered out 0
>> 933  607.9826    3    null   0.0      filtered out 0
>> 934  607.9816    3    null   0.0      filtered out 0
>> 935  401.2254    0    null   0.0      filtered out 0
>> 936  877.4576    3    null   0.0      filtered out 0
>> 937  494.8108    2    null   0.0      filtered out 0
>> 938  494.8096    0    null   0.0      filtered out 0
>> 939  494.8149    2    null   0.0      filtered out 0
>> 940  494.8087    0    null   0.0      filtered out 0
>> 941  494.8096    0    null   0.0      filtered out 0
Qualified spectra: 10
Total spectra: 942
```

The output file should look like the following:

```
Reconstruction  Score   Accuracy
[229.11525]IQIV[390.19635][340.17102]VPK        99.10   0.10 %
[342.17212]QIV[390.19635][212.10669]QVPK        98.86   0.10 %
[229.11525]IQIV[602.30304]QVPK  97.97   0.10 %
[342.17212]QIV[390.19635]V[241.12122]VPK        97.59   2.96 %
[243.12228]VQIV[390.19635][340.17102]VPK        97.48   2.58 %
[341.1716]EIV[390.19635][212.10669]QVPK 97.35   2.48 %
[229.11525]IQIV[489.24615][241.12122]VPK        96.71   2.60 %
```

3.2 Protein Identification

1. Discard any peptide less than seven amino acids residues. Filter out any low scoring peptides from the de novo peptide lists generated from the previous step (refer to [13] for MS data validation techniques).
2. Compare the list generated from **step 1** with the proteins identified by Mascot search (described in Subheading 2.1.1 [3]) and create a nonredundant de novo peptide list.
3. Convert the peptides to FASTA format (*see* **Note 7**).
4. Download and install the latest BLAST [18] package on your machine.
5. Make the BLAST [18] databases with the protein sequences mentioned in Subheading 2.1.1 [7]. The command to create a BLAST database is:

```
makeblastdb -in <inputfile> -out <output-
file> -dbtype prot
```

6. Perform a BLASTP-short search task against the protein database created in the previous step, using default parameters with a minimum E-value of 20000, filter (seg) = off, word size = 2, composition based statistics = off, score matrix = PAM30, no gaps to identify homologous proteins to target sequences. Sample command:

```
blastp -task blastp-short -num_threads <n>
-query <input FASTA> -db <path to blast
database> -out <output file> -evalue 20000
-outfmt 6 -comp_based_stats 0 -ungapped
-matrix PAM30 -seg no -word_size 2
```

7. Only consider sequences with 100 % identity and coverage for protein identification (*see* **Note 8**).

3.3 Sequential-BLAST Similarity Search and Functional Annotation

1. Sequential Database similarity search.

   Sequential BLAST refers to running BLAST repeatedly [26] against specific databases, filtering off the most reliable hits from each run. Refer to the protocol set out in [13] and perform a database similarity search using the databases listed in Subheading 2.1.2 in the order in which they appear. Use the protein sequences listed in Subheading 2.1.1 [7] as the input for this step.
2. Protein functional domains and motifs, and Gene Ontology [15].

   Refer to the steps in [13] and perform functional annotations using the protein sequences from Subheading 2.1.1 [7].

4 Notes

1. We have used the raw files of nanoLC-ESI-MS/MS data from our previous study [15], 11 fractions were converted to .mgf format. This data was then submitted to Mascot and searched against *T. melanosporum* database. The protein database comprises 12771 nonredundant sequences from truffle proteome [14]. The parameters used include MS tolerance of ±100 ppm and MS/MS tolerance of ±0.2 Da. The fixed modification was set to carbamidomethylation and modification of methionine, threonine and deamination of asparagine and glutamine were included as variable modifications. These Mascot results were compared with the de novo sequencing results obtained in the current study. This method can be used for any MS/MS data. A wide range of open source tools [27] are available to convert MS data to other compatible formats.
2. You can click the *Running Info* Tab on the bottom left hand corner, to view the progress.
3. You can examine the peptides and their spectra individually. Amino acids are color-coded based on their confidence level. Red (>90%) is very good score, purple (80%-90%) is good while blue (60 %–80 %) is considered to be acceptable score. You can also adjust the local confidence threshold to filter the de novo sequence with desired/highly confident sequence tags. You can hover your mouse over a peptide and examine the confidence value of each amino acid. However for the entire peptide sequence, it is recommended to use the expected percentage of correct amino acids/average local confidence (ALC) value.
4. To process large amounts of files use –list option to give a list of input files otherwise the program will reread models for each input file. You can easily create a file with list of MS files within a folder from a Linux shell. To do this, change directory to the folder that contains the raw files, then run the following command

```
ls -d -1 $PWD/*.* >file-list.txt
```

This will write the full path of all available files to *file-list.txt* file. You can now pass the *file-list.txt* file with –list option. Sample command run PepNovo using a file list:

```
./PepNovo_bin -list </full/path/to/
file_list> -model CID_IT_TRYP -PTMs
C+57:M+16:N+1 -fragment_tolerance 0.03
-pm_tolerance 0.05 -digest TRYPSIN >
<path/to/output/file>
```

5. You need to have java runtime environment (JRE) 1.6 or greater installed on your machine to run UniNovo. To optimize the performance, you can set the maximum Java heap size (i.e., run it with -Xmx<size> option) to allow UniNovo to use a maximum amount of memory. For example you could run the same command stated in step 2 of 3.1.3 with the following command to allocate 2000MB memory.

```
java -jar -Xmx2000m UniNovo.jar -i <input
file > -o <output file prefix> -l 10 -acc 0.6
-t 0.03Da -pt 0.05Da -f CID -e 1
```

6. If you are processing a large number of files, it is best to write a simple script (bash, Python, Perl) to process them in a batch. Here is a basic script written in python that takes the source and destination directory as an input and process all files sequentially.

```
#!/usr/bin/env python
#Filename: run_uninovo.py
#sample command: python run_uninovo.py
<full path/of/the/directory containing MS
files>
#run this program from the UniNovo directory
import os, sys, shutil, psutil
import subprocess as sp
def run_process(cmd, logfile):
    """ execute a process"""
    p = sp.Popen(cmd, shell=True,
        stdout=sp.PIPE, stderr=sp.STDOUT
    stdout = []
    while True:
        line = p.stdout.readline()
        stdout.append(line)

        if line == '' and p.poll() != None:
            break

    loglist=''.join(stdout)
    outfile=open(logfile, "w")
    outfile.write(loglist)
    outfile.close()
    return p.returncode

def run_uninovo(infile,outfile):
    print "processing..",infile
    cmd="java -jar UniNovo.jar -i
    "+infile+" -o "+outfile +" -l 10 -acc
    0.6 -t 0.03Da -pt 0.05Da -f CID -e 1"
    logfile=outfile+".log"
    code=run_process(cmd, logfile)
```

```
    success=""
    if code<>0:
     success=-1
     print "Failed to run UniNovo
     for",infile, "please check the log
     file"
    else:
     success=1
    return success

 if __name__ == "__main__":
    directory=sys.argv[1]
    retrn_code=0
    for root, dirs, files in
    os.walk(directory):
        for file in files:
            if file.endswith('.mgf') or file.
            endswith('.mzXML') or
  file.endswith('.ms2'):
                out_string=os.path.
splitext(file)[0]+"_uninovo_out"
                retrn_code=run_
uninovo(directory+"/"+file, out_string)
```

7. Peptides must be converted to FASTA format for BLAST [18] alignment. A simple script can be written to achieve this. Below is an example of a basic python script that can create FASTA files from an excel sheet that contains one peptide per cell. It will also exclude all peptides less than the minimum length.

```
#!/usr/bin/env python
#Filename: generate_pep_fasta.1.0.py
#sample command: python generate_pep_
fasta.1.0.py <path to excel file with
peptide> <output file name> <sequence id
string> <minimum peptide length>
import sys, string, os
from xlrd import open_workbook
def
generate__fasta_seq(infilename,outfilename,s
ample_name,min_pep_length):
        f = open(outfilename,'w')
        book = open_workbook(infilename)
        sheet0 = book.sheet_by_index(0)
        count=0
        skipped=0
        sample=sample_name
        seq_no=1
        data=[]
        data=sheet0.col(0)
```

```
        for i in data:
                if len(i.value)>= int(min_
                pep_length):
                        text=
">PEP|"+sample+"|MMB"+str(seq_no).
zfill(6)+'\n'
                        f.write(text)
                        pep=i.value+'\n'
                        f.write(pep)
                        count+=1
                        seq_no+=1
                else:
                        skipped+=1
                        print len(i.value)
        print count, "peptides are added to
        ", outfilename
        print skipped, "peptides are less
        than the minimum length",min_pep_
        length
        f.close()
        if __name__ == '__main__':
    infilename=sys.argv[1]
    outfilename=sys.argv[2]
    sample_name=sys.argv[3]
    min_pep_length=sys.argv[4]
    generate__fasta_seq(infilename,outfilename
    ,sample_name,min_pep_length)
```

8. A protein is identified if two proteotypic peptides with minimum 7 residues, as accepted by the proteomics community at large. Alternatively a single peptide of 9 amino acids can also be considered sufficient for identification [28]

References

1. Eng JK, McCormack AL, Yates JR (1994) An approach to correlate tandem mass spectral data of peptides with amino acid sequences in a protein database. J Am Soc Mass Spectrom 5(11):976–989. doi:10.1016/1044-0305(94)80016-2
2. Turetschek R, Lyon D, Desalegn G, Kaul HP, Wienkoop S (2016) A proteomic workflow using high-throughput de novo sequencing towards complementation of genome information for improved comparative crop science. Methods Mol Biol 1394:233–243. doi:10.1007/978-1-4939-3341-9_17
3. Devabhaktuni A, Elias JE (2016) Application of de novo sequencing to large-scale complex proteomics data sets. J Proteome Res. doi:10.1021/acs.jproteome.5b00861
4. Biemann K, Cone C, Webster BR, Arsenault GP (1966) Determination of the amino acid sequence in oligopeptides by computer interpretation of their high-resolution mass spectra. J Am Chem Soc 88(23):5598–5606
5. Seidler J, Zinn N, Boehm ME, Lehmann WD (2010) De novo sequencing of peptides by MS/MS. Proteomics 10(4):634–649. doi:10.1002/pmic.200900459
6. Vyatkina K, Wu S, Dekker LJ, VanDuijn MM, Liu X, Tolic N, Dvorkin M, Alexandrova S, Luider TM, Pasa-Tolic L, Pevzner PA (2015) De novo sequencing of peptides from top-down tandem mass spectra. J Proteome Res 14(11):4450–4462. doi:10.1021/pr501244v

7. Carvalho PC, Lima DB, Leprevost FV, Santos MD, Fischer JS, Aquino PF, Moresco JJ, Yates JR 3rd, Barbosa VC (2016) Integrated analysis of shotgun proteomic data with PatternLab for proteomics 4.0. Nat Protoc 11(1):102–117. doi:10.1038/nprot.2015.133
8. Taylor JA, Johnson RS (1997) Sequence database searches via de novo peptide sequencing by tandem mass spectrometry. Rapid Commun Mass Spectrom 11(9):1067–1075. doi:10.1002/(SICI)1097-0231(19970615)11:9<1067::AID-RCM953>3.0.CO;2-L
9. Frank A, Pevzner P (2005) PepNovo: de novo peptide sequencing via probabilistic network modeling. Anal Chem 77(4):964–973
10. Ma B, Zhang K, Hendrie C, Liang C, Li M, Doherty-Kirby A, Lajoie G (2003) PEAKS: powerful software for peptide de novo sequencing by tandem mass spectrometry. Rapid Commun Mass Spectrom 17(20):2337–2342. doi:10.1002/rcm.1196
11. Fischer B, Roth V, Roos F, Grossmann J, Baginsky S, Widmayer P, Gruissem W, Buhmann JM (2005) NovoHMM: a hidden Markov model for de novo peptide sequencing. Anal Chem 77(22):7265–7273. doi:10.1021/ac0508853
12. Jeong K, Kim S, Pevzner PA (2013) UniNovo: a universal tool for de novo peptide sequencing. Bioinformatics 29(16):1953–1962. doi:10.1093/bioinformatics/btt338
13. Islam MT, Mohamedali A, Nawar I, Baker MS, Ranganathan S (2016) A systematic bioinformatics approach to identify high quality MS data and functionally annotate proteins and proteomes. Proteome Bioinformatics, Methods in molecular biology. Springer, New York, NY. doi: 10.1007/978-1-4939-6740-7
14. Martin F, Kohler A, Murat C, Balestrini R, Coutinho PM, Jaillon O, Montanini B, Morin E, Noel B, Percudani R, Porcel B, Rubini A, Amicucci A, Amselem J, Anthouard V, Arcioni S, Artiguenave F, Aury JM, Ballario P, Bolchi A, Brenna A, Brun A, Buee M, Cantarel B, Chevalier G, Couloux A, Da Silva C, Denoeud F, Duplessis S, Ghignone S, Hilselberger B, Iotti M, Marcais B, Mello A, Miranda M, Pacioni G, Quesneville H, Riccioni C, Ruotolo R, Splivallo R, Stocchi V, Tisserant E, Viscomi AR, Zambonelli A, Zampieri E, Henrissat B, Lebrun MH, Paolocci F, Bonfante P, Ottonello S, Wincker P (2010) Perigord black truffle genome uncovers evolutionary origins and mechanisms of symbiosis. Nature 464(7291):1033–1038. doi:10.1038/nature08867
15. Islam MT, Mohamedali A, Garg G, Khan JM, Gorse AD, Parsons J, Marshall P, Ranganathan S, Baker MS (2013) Unlocking the puzzling biology of the black Perigord truffle Tuber melanosporum. J Proteome Res 12(12):5349–5356. doi:10.1021/pr400650c
16. UniProt Consortium (2015) UniProt: a hub for protein information. Nucleic Acids Res. 43 (Database issue):D204–212. doi: 10.1093/nar/gku989.
17. Berman HM, Westbrook J, Feng Z, Gilliland G, Bhat TN, Weissig H, Shindyalov IN, Bourne PE (2000) The Protein Data Bank. Nucleic Acids Res 28(1):235–242
18. Altschul SF, Gish W, Miller W, Myers EW, Lipman DJ (1990) Basic local alignment search tool. J Mol Biol 215(3):403–410. doi:10.1016/S0022-2836(05)80360-2
19. NCBI BLAST ftp://ftp.ncbi.nlm.nih.gov/blast/executables/blast+/LATEST/. Accessed 26 October 2016
20. Quevillon E, Silventoinen V, Pillai S, Harte N, Mulder N, Apweiler R, Lopez R (2005) InterProScan: protein domains identifier. Nucleic Acids Res 33(Web Server issue):116–120. doi:10.1093/nar/gki442
21. InterProScan . http://www.ebi.ac.uk/interpro/search/sequence-search. Accessed 26 October 2016
22. Moriya Y, Itoh M, Okuda S, Yoshizawa AC, Kanehisa M (2007) KAAS: an automatic genome annotation and pathway reconstruction server. Nucleic Acids Res 35(Web Server issue):182–185. doi:10.1093/nar/gkm321
23. Xie C, Mao X, Huang J, Ding Y, Wu J, Dong S, Kong L, Gao G, Li CY, Wei L (2011) KOBAS 2.0: a web server for annotation and identification of enriched pathways and diseases. Nucleic Acids Res 39(Web Server issue):316–322. doi:10.1093/nar/gkr483
24. Allmer J (2011) Algorithms for the de novo sequencing of peptides from tandem mass spectra. Expert Rev Proteomics 8(5):645–657. doi:10.1586/epr.11.54
25. Steen H, Mann M (2004) The ABC's (and XYZ's) of peptide sequencing. Nat Rev Mol Cell Biol 5(9):699–711. doi:10.1038/nrm1468
26. Ranganathan S, Khan JM, Garg G, Baker MS (2013) Functional annotation of the human chromosome 7 "missing" proteins: a bioinformatics approach. J Proteome Res 12(6):2504–2510. doi: 10.1021/pr301082p.
27. Mass spectrometry data format. https://en.wikipedia.org/wiki/Mass_spectrometry_data_format. Accessed on 26 October 2016
28. Omenn GS, Lane L, Lundberg EK, Beavis RC, Nesvizhskii AI, Deutsch EW (2015) Metrics for the Human Proteome Project 2015: progress on the human proteome and guidelines for high-confidence protein identification. J Proteome Res 14(9):3452–3460. doi:10.1021/acs.jproteome.5b00499

Chapter 11

Phylogenetic Analysis Using Protein Mass Spectrometry

Shiyong Ma, Kevin M. Downard, and Jason W.H. Wong

Abstract

Through advances in molecular biology, comparative analysis of DNA sequences is currently the cornerstone in the study of molecular evolution and phylogenetics. Nevertheless, protein mass spectrometry offers some unique opportunities to enable phylogenetic analyses in organisms where DNA may be difficult or costly to obtain. To date, the methods of phylogenetic analysis using protein mass spectrometry can be classified into three categories: (1) de novo protein sequencing followed by classical phylogenetic reconstruction, (2) direct phylogenetic reconstruction using proteolytic peptide mass maps, and (3) mapping of mass spectral data onto classical phylogenetic trees. In this chapter, we provide a brief description of the three methods and the protocol for each method along with relevant tools and algorithms.

Key words Phylogenetics, De novo sequencing, Mass mapping, Molecular evolution, Phylogenetic tree, Mass tree

1 Introduction

In biology, phylogenetic analyses are used to assess the evolutionary relationships among a set of organisms. To do so, the degree of homology among the features of the taxa needs to be measured and compared. Traditionally, homology measurements have typically been based on morphological data, which is usually acquired by measuring certain phenotypes of the investigated organisms. With the emergence of molecular biology and technological advances in DNA sequencing, molecular phylogenetic has now become the standard method for phylogenetic analysis [1].

Molecular phylogenetics typically uses DNA or translated DNA sequences of specific genes to measure the evolutionary distance across species, populations, or homologous genes within an individual organism. DNA sequences are generally acquired after its amplification using the polymerase chain reaction followed by Sanger [2] or massive-parallel sequencing [3]. While this process is robust and well established, there can be circumstances that limit the ability to sequence DNA for phylogenetic analysis. For instance,

Shivakumar Keerthikumar and Suresh Mathivanan (eds.), *Proteome Bioinformatics*, Methods in Molecular Biology, vol. 1549, DOI 10.1007/978-1-4939-6740-7_11, © Springer Science+Business Media LLC 2017

phylogenetic analysis of fossilized animals can be challenging due to DNA degradation [4, 5]. Furthermore, although massive-parallel sequencing is becoming increasingly affordable, when the sequencing of a large number of samples is required, cost and throughput can still be limiting factors [6].

Mass spectrometry has been demonstrated to provide an alternative method for phylogenetic analysis through the acquisition of protein rather than genetic sequence information [4, 5, 7–9]. Tandem mass spectrometry-based de novo sequencing has been used to elucidate protein sequences from fossilized tissue for phylogenetic reconstruction [4, 5]. The use of proteolytic peptide masses from mass map or fingerprint data, without peptide sequencing, using the so-called "Mass Trees" has also been shown to be applicable for the phylogenetic reconstruction and analysis of strains of the influenza virus [7, 9]. Furthermore, it has recently been demonstrated that using the FluClass algorithm, phylogenetic classification of the influenza virus can be performed by directly mapping peptide mass fingerprints onto existing phylogenetic trees using mass, as opposed to sequence, information [8].

The use of de novo sequencing to generate protein sequence is akin to traditional protein-based molecular phylogenetic techniques including sequence generation using Edman degradation [10] as well as biochemical and biophysical methods such as using immunological information [11] or measuring protein electrophoretic properties [12]. Nevertheless, the sensitivity of modern mass spectrometers provides significant advantages over those traditional methods particularly when the analysis needs to be performed on highly degraded samples.

The MassTree algorithm [6] reads sets (M_t) of monoisotopic m/z values m_1, m_2, ..., mn for peptide ions calculated theoretically, or detected in a mass spectrum, following the proteolytic digestion of the viral protein. Proteins are grouped to a clade or subclade based on the number of m/z values of the total contained within sets that are indistinguishable i by mass; that is, the difference between them is less than a specified mass error (default is 5 ppm). A distance score between two sets of masses (M_1 and M_2) is then computed based on the number of matching mass values within each set. The relative length of the branches reflects the proportion of matching peptide masses of the total. Among the peptide ions detected, pairs of mass values from different mass sets (M_t) that differ in mass which correspond to a single amino acid substitution, s, are also determined and weighted.

The FluClass algorithm [8], reads in a phylogenetic tree which can be generated using sequence-based methods. It then generates a list of mass corresponding to theoretical peptide ion for all nodes of the tree including internal nodes. The algorithm then reads in a list of monoisotopic peptide ion masses before using a novel scoring method to identify the node most phylogenetically related to the input mass spectrum.

In the case of MassTree and FluClass algorithms, the methods leverage the high throughput capabilities and speed of analysis of mass spectrometry. These are of particular benefit where the processing and sequencing of DNA for a large numbers of samples proves challenging. Although these methods have been applied in published reports to the analysis of influenza viral proteins [13–16], they can be applied to study the phylogenetic relationship between orthologous proteins across different organisms or paralogous proteins within a single organism. This chapter outlines the steps required to facilitate phylogenetic analysis using protein mass spectrometry data using each of the three methods described.

2 Materials

This chapter assumes that all methods for the acquisition of experimental mass spectral data are already available to the reader. It therefore is focused on the methods for computational phylogenetic-like analysis only. Below is a list of free-for-academic-use software required to conduct the data analyses described in this chapter:

- PepNovo+ [17] (http://proteomics.ucsd.edu/Software/PepNovo/)—for *de novo* sequencing of tandem mass spectral data.
- FluClass [8] (https://powcs.med.unsw.edu.au/fluclass)—for scoring of peptide mass mapping mass spectral data against existing phylogenetic tree.
- Archaeopteryx [18] (https://sites.google.com/site/cmzmasek/home/software/archaeopteryx)—for the visualization of phylogenetic trees.
- Proteowizard [19] (http://proteowizard.sourceforge.net/)—for converting mass spectral data file formats for vendor specific formats to standard mass spectrometry data formats compatible with downstream data analysis pipelines such as mgf, dta, mzXML, and mzML.

Before proceeding, the reader should ensure that the software are installed and functional per the instructions provided for each software package.

3 Methods

This chapter is focused on the phylogenetic analysis using mass spectral data for a set of homologous proteins. A schematic diagram outlining the three methods is shown in Fig. 1.

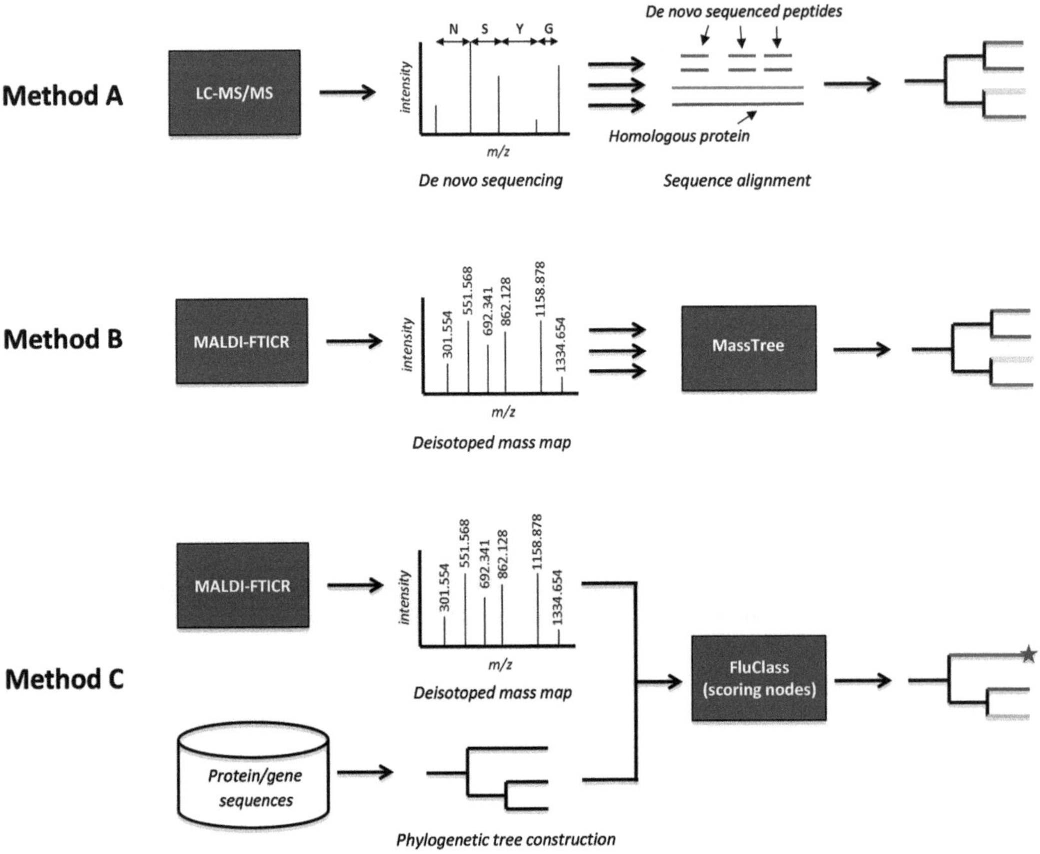

Fig. 1 Schematic illustration of the three methods for phylogenetic analysis using protein mass spectrometry data. *Method A*—*De novo* protein sequencing followed by classical phylogenetic reconstruction. *Method B*—Direct phylogenetic reconstruction using proteolytic peptide mass maps. *Method C*—Mapping of mass spectral data onto classical phylogenetic trees

3.1 De Novo Protein Sequencing Followed by Classical Phylogenetic Reconstruction

1. Acquire tandem mass spectral data from the tryptic digest of a protein sample. LC-MS/MS data for a tryptic digest of a purified protein is recommended for optimal sequence coverage.
2. Use Proteowizard's MSConvert tool [19] to convert the raw mass spectral data format to mgf format. The mgf format is a text-based format introduced by the Mascot software [20] and is used to store all tandem mass spectra across LC-MS/MS experiment. As mass spectra of precursor ions (i.e., MS1 level data) are not necessary for this analysis, we recommend mgf format because of the smaller file size (*see* **Note 1**).
3. Use PepNovo+ [17] to determine candidate peptide sequences from each MS/MS spectrum. An example command for PepNovo+ is the following: *./PepNovo_bin -file test.mgf -model CID_IT_TRYP -PTMs C+57 -digest TRYPSIN -min_filter_prob 0.9>output.txt*. For a comprehensive description

of each of the parameters, refer to the PepNovo readme file (*see* **Note 2**).

4. Map each de novo sequenced fragment to the homologous sequence using the Smith-Waterman pairwise alignment algorithm [21]. An online implementation of the algorithm can be found at the EBI web site (http://www.ebi.ac.uk/Tools/psa/emboss_water/index.html) (*see* **Note 3**). The distance similarity matrix should be chosen according to the expected sequence similarity with the homologous protein sequence (*see* **Note 4**). *De novo* sequences that do not align to the homologous protein sequence should be disregarded from further analysis.
5. Using the combined output from the Smith-Waterman alignment of all *de novo* sequenced peptides, the sequence of the de novo sequenced protein can be reconstructed. **Steps 1–4** should be repeated for each protein being analyzed.
6. To begin phylogenetic analysis, the aligned sequences (in fasta format) containing any homologous protein sequences from the protein databases, such as UniProt [22], and each *de novo* sequenced protein are then constructed. For this analysis, any region that is not covered by *de novo* sequencing is left as gaps, i.e., denoted by '-' (*see* **Note 5**).
7. Once the sequences are aligned in fasta format, phylogenetic reconstruction methods can be applied. There is a diverse range of methods for phylogenetic reconstruction (*see* **Note 6**). For simplicity, and as an illustrative example, the distance-based neighbor-joining method from the EBI ClustalW2 [23] package can be used with this protocol (http://www.ebi.ac.uk/Tools/phylogeny/clustalw2_phylogeny/). The excluded gaps parameter can be selected in the ClustalW2 algorithm to ensure that only regions *de novo* sequenced across all samples are used for phylogenetic analysis. Select Newick/PHYLIP as the output tree format.
8. The phylogenetic tree can now be visualized using a tree visualization tool such as Archaeopteryx [18].

3.2 Direct Phylogenetic Reconstruction Using Proteolytic Peptide Mass Maps

1. Acquire mass map data (i.e., precursor ion spectra) on the tryptic digest of a protein sample. High resolution MALDI-FTICR data is recommended as high mass accuracy is preferable, while the use of MALDI ensures the proteolytic fragment masses are detected as singly charged ions, thus removing the need for spectral deconvolution to obtain peptide molecular weights.
2. Generate a mass list containing deisotoped or monoisotopic masses of all peptide ions detected (*see* **Note 7**).
3. Repeat **steps 1** and **2** across all samples.

Table 1
Example input file format for MassTree

HA_Cambodia	842.5463	871.5404	921.4428	1152.547	1168.542	1175.571	1299.477	...
Duck_Hunan	529.3471	625.3788	643.3212	842.541	871.541	886.5129	893.4743	...
HA_HK_2003	886.5119	903.4327	921.4439	1152.548	1168.543	1175.57	1185.436	...
HA_Indonesia	842.5458	1152.547	1168.542	1175.57	1313.493	1329.49	1500.66	...
Magpie_2003	842.5466	871.5409	886.5112	903.4334	921.4435	1152.547	1168.543	...
Duck_NZL	780.429	887.5351	900.4899	1096.545	1112.539	1172.627	1203.575	...
Swine_Guangx	885.5559	900.4892	917.4477	935.4498	935.4586	1112.54	1186.695	...
HA_HK_97	842.5439	871.5376	1123.68	1314.559	1510.873	2156.09	2196.957	...
HA_Turkey	842.5461	871.5392	921.4424	1152.547	1168.542	1175.571	1299.477	...
HA_Vietnam	842.5462	871.5409	921.4429	1152.548	1168.542	1175.571	1299.478	...
SolomonIs_2006	780.429	827.4349	827.4407	944.5562	945.4533	1250.601	1268.612	...
California_2009	780.4302	885.5565	887.4087	903.4038	916.4533	1054.554	1109.559	...

The first column contains the protein name and in each row, the associated mass list of each protein is on the subsequent columns

4. Prepare an input file for MassTree [7]. This file needs to be in tab delimited format with each row containing the sample name in the first column followed by the mass in subsequent columns sorted in ascending order according to the mass values. *See* Table 1 for example.
5. Run the MassTree algorithm [7] from the following web site. http://flu.med.unsw.edu.au/kdownard/MassTree_v2.html. This is achieved by simply uploading the input file and clicking the "Submit" button.
6. Once the algorithm is complete, the output file in Newick/PHYLIP format can be downloaded and visualized using Archaeopteryx [18]. For an example of a MassTree in comparison to a sequence-based tree, *see* Fig. 2 and **Note 8**).

3.3 Mapping of Mass Spectral Data onto Classical Phylogenetic Trees

1. Acquire gene/protein sequences in fasta format for phylogenetic tree reconstruction (*see* **Note 9**).
2. Align the protein sequences using the ClustalΩ algorithm [24] (at http://www.ebi.ac.uk/Tools/msa/clustalo/). For the alignment of gene sequences, use MUSCLE algortihm [25] (at http://www.ebi.ac.uk/Tools/msa/muscle/). In either case, output the alignment file in Pearson/FASTA format (*see* **Note 10**).
3. Proceed to phylogenetic tree reconstruction using the neighbor-joining method provided within the EBI

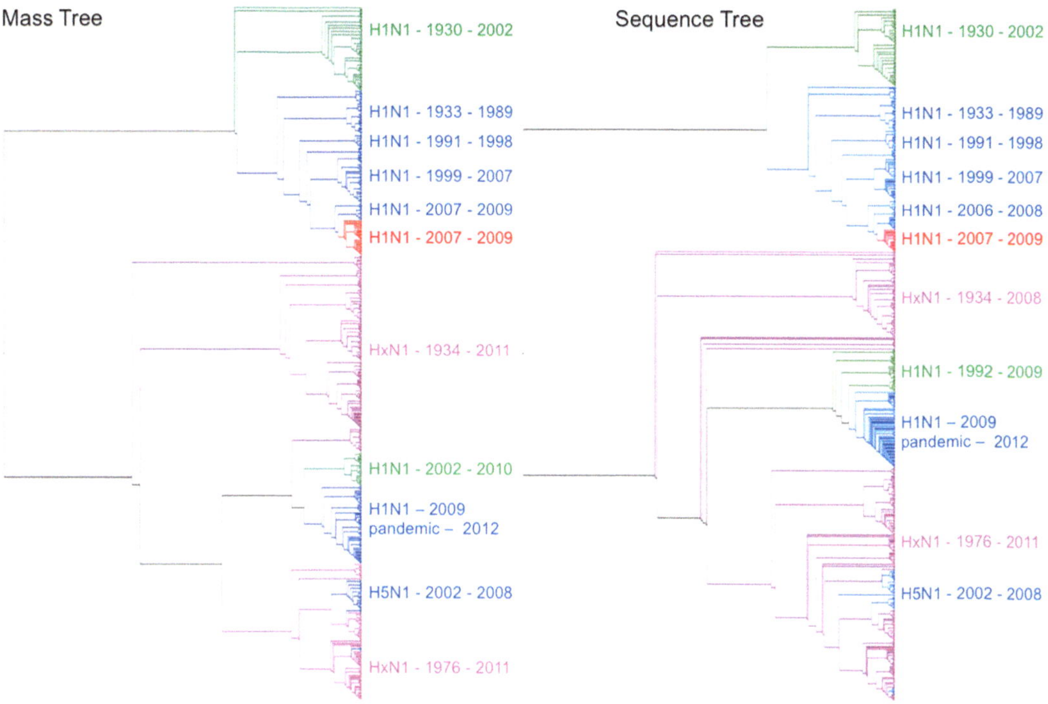

Fig. 2 Example of a MassTree in comparison to a sequence-based. Mass (*left*) and sequence (*right*) trees constructed for a subset of 595 N1 neuraminidase sequences of human HxN1 virus strains. Reprinted with permission from ref. 8. Copyright 2016 American Chemical Society

ClustalW2 [23] package (http://www.ebi.ac.uk/Tools/phylogeny/clustalw2_phylogeny/) (*see* **Note 6**). Save the output file in Newick/PHYLIP format.

4. Acquire mass map data from the tryptic digest of a protein sample. High resolution MALDI-FTICR data is recommended for the accurate identification of proteolytic fragment masses from singly charged ions.
5. Generate a mass list containing the deisotoped or monoisotopic masses of all peptide ions detected (*see* **Note 7**).
6. Use FluClass [8] to score the input mass spectrum against nodes of the phylogenetic tree. An example input command is as follows: *FluClass.exe –infasta example.fa –intree example.phy –inmass example_spectrum.txt*.
7. FluClass will output a tab delimited file containing scores of the input spectrum against each of the nodes of the input phylogenetic tree. FluClass also outputs colorized phylogenetic trees based on the scores in Newick/PHYLIP format. These trees can be visualized using Archaeopteryx [18]. *See* Fig. 3 for example colorized tree generated by FluClass.

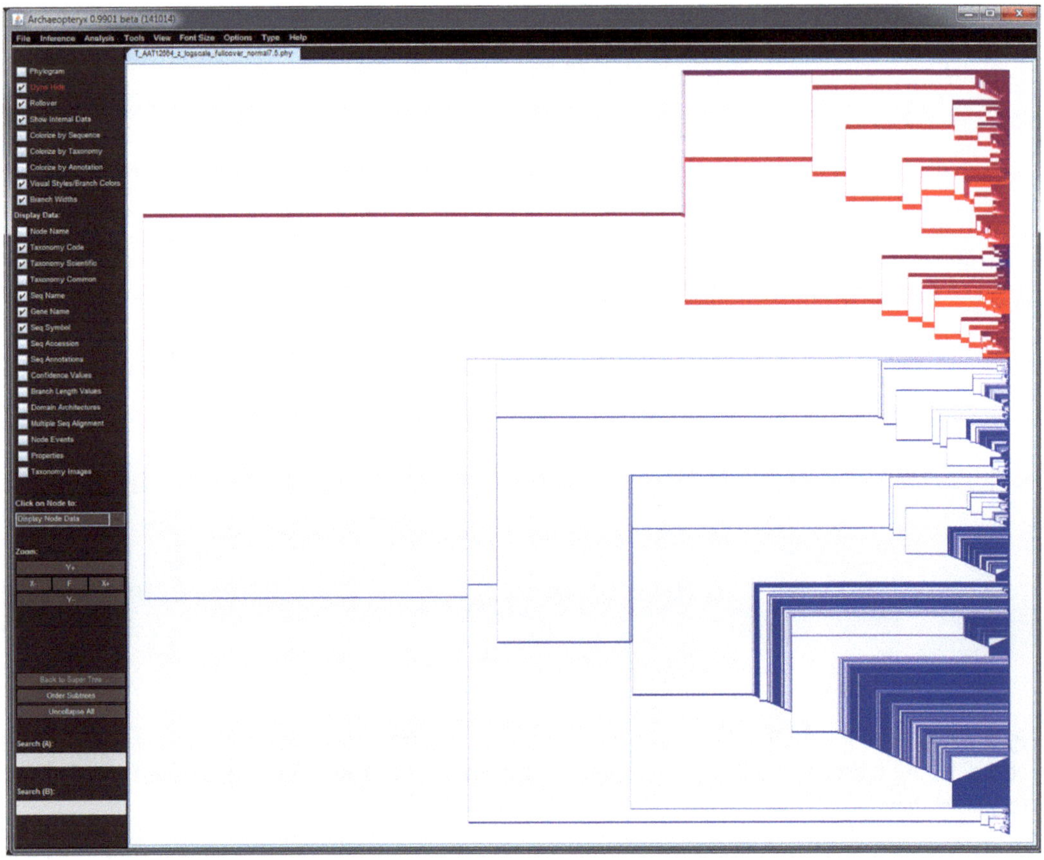

Fig. 3 Colorized tree generated by FluClass visualized in Archaeopteryx. The tree was generated using all human hosted influenza type A hemagglutinin (HA). The branches colored *red* are those where the mass spectrum matches most closely with the associated node while the *blue* colored branches are those that are the lowest scoring

4 Notes

1. Take note of the type of MS/MS data acquired in the experiment. For low resolution collision-induced dissociation (CID) MS/MS data, default parameters from MSConvert can be used. However, for high resolution higher energy collisional dissociation (HCD) data, ensure that peak picking is selected such that only monoisotopic ions are retained in order to avoid an excessive number of peaks in the resulting mgf file.
2. De novo sequencing of peptides using MS/MS data can be challenging and is not foolproof. To avoid false sequences, it is recommended that PepNovo+ [17] be first used on known protein/peptides analyzed using the same mass spectrometer such that parameters and scoring filters can appropriately

adjusted for a particular dataset. It is also recommended that, where possible, MS/MS spectra be manually validated to ensure that the correct sequences are being used for downstream data analysis.

3. Where no homologous sequences are available, it is necessary to perform multiple LC-MS/MS analyses proteins digested using two or more different endoproteinases with complementary cleavage site specificity. This enables peptides with overlapping sequences to be obtained such that their assembly in the correct order within the de novo protein sequence can be performed [26].
4. When examining proteins from species that are closely related phylogenetically, higher similarity substitution matrices such as BLOSUM80 or BLOSUM90 can be used [27]. The default BLOSUM62 matrix is most commonly used as it provides a balanced substitution frequency that is suitable for most analyses.
5. To ensure that each de novo sequence is compared with equal phylogenetic weighting, only the sequences of regions that have been de novo sequenced across all samples should be retained. Certain phylogenetic analysis tools such as ClustalW2 [23] automatically exclude gapped regions in any sequence for use in phylogenetic analyses. However, depending on the downstream tool, masking of these regions, including those found in the sequences of homologous proteins within databases needs to be performed. This can be achieved by either replacing those amino acids with lower case letters or replacing them with the symbol 'X'.
6. Molecular phylogenetics methods can broadly be classified as either distance-based, maximum parsimony, maximum likelihood and Bayesian methods. For a comprehensive review and comparison of these methods, refer to references [28, 29].
7. It is important that only monoisotopic masses obtained from a mass spectrum are selected for phylogenetic analysis. Vendor software, or an open access tool such as Hardklor [30], should be used in order to achieve this.
8. Figure 2 shows the mass and phylogenetic (gene sequence) trees for full-length H1 human influenza using one strain from each country for each available year through 2012. Note that the mass tree on the left is built solely from sets of masses for proteolytic peptide segments of influenza H1 hemagglutinin. It is clear that both trees show very similar topologies, with common-colored branches containing strains that are identical on both trees. A comparison of sets of trees, using two different tree comparison algorithms, showed that they were 60.1% congruent, a value not dissimilar to those obtained when two

sequence trees constructed with two different tree building algorithms are compared [7].

9. It is generally advantageous to use DNA sequences for closely related species as DNA sequence contains more evolutionary information than protein sequences. However, for more distantly related sequences, protein sequences generally allows for more accurate phylogenetic inference [31]. There are also advantages of using protein sequences over gene sequences especially for the study of microorganisms [32]. The main reason is due to the degeneracy of the genetic code. All but two amino acids (Met and Trp) are encoded by at least two codons. Most changes in the third codon position do not affect the encoded protein sequence. Proteins have 20 possible amino acids at each position (vs. 4 for DNA/RNA) that provides a stronger phylogenetic signal [32].
10. A range of multiple sequence alignment algorithms are available for the alignment of both DNA and protein sequences. For a comprehensive review, *see* [33].

Acknowledgments

Part of the results presented and cited in this chapter were supported with funds from an Australian Research Council Discovery Project Grant (DP120101167) awarded to K.M.D. and J.W.H.W. S.M. is supported by a China Scholarship Council and UNSW Australia Tuition Fee scholarships. J.W.H.W. is supported by an Australian Research Council Future Fellowship (FT130100096).

References

1. Kimura M (1983) The neutral theory of molecular evolution. Cambridge University Press, Cambridge
2. Sanger F, Coulson AR (1975) A rapid method for determining sequences in DNA by primed synthesis with DNA polymerase. J Mol Biol 94(3):441–448
3. Mardis ER (2008) The impact of next-generation sequencing technology on genetics. Trends Genet 24(3):133–141. doi:10.1016/j.tig.2007.12.007
4. Asara JM, Schweitzer MH, Freimark LM, Phillips M, Cantley LC (2007) Protein sequences from mastodon and Tyrannosaurus rex revealed by mass spectrometry. Science 316(5822):280–285. doi:10.1126/science.1137614
5. Cappellini E, Jensen LJ, Szklarczyk D, Ginolhac A, da Fonseca RA, Stafford TW, Holen SR, Collins MJ, Orlando L, Willerslev E, Gilbert MT, Olsen JV (2012) Proteomic analysis of a pleistocene mammoth femur reveals more than one hundred ancient bone proteins. J Proteome Res 11(2):917–926. doi:10.1021/pr200721u
6. Sboner A, Mu XJ, Greenbaum D, Auerbach RK, Gerstein MB (2011) The real cost of sequencing: higher than you think! Genome Biol 12(8):125. doi:10.1186/gb-2011-12-8-125
7. Lun AT, Swaminathan K, Wong JW, Downard KM (2013) Mass trees: a new phylogenetic approach and algorithm to chart evolutionary history with mass spectrometry. Anal Chem 85(11):5475–5482. doi:10.1021/ac4005875
8. Ma S, Downard KM, Wong JW (2015) FluClass: a novel algorithm and approach to score and visualize the phylogeny of the influenza virus using mass spectrometry. Anal Chim Acta 895:54–61. doi:10.1016/j.aca.2015.09.004

9. Swaminathan K, Downard KM (2014) Evolution of influenza neuraminidase and the detection of antiviral resistant strains using mass trees. Anal Chem 86(1):629–637. doi:10.1021/ac402892m
10. Edman P (1949) A method for the determination of amino acid sequence in peptides. Arch Biochem 22(3):475
11. Prager EM, Welling GW, Wilson AC (1978) Comparison of various immunological methods for distinguishing among mammalian pancreatic ribonucleases of known amino acid sequence. J Mol Evol 10(4):293–307
12. Harris H (1966) Enzyme polymorphisms in man. Proc R Soc Lond B Biol Sci 164(995):298–310
13. Downard KM (2013) Proteotyping for the rapid identification of influenza virus and other biopathogens. Chem Soc Rev 42(22):8584–8595. doi:10.1039/c3cs60081e
14. Lun AT, Wong JW, Downard KM (2012) FluShuffle and FluResort: new algorithms to identify reassorted strains of the influenza virus by mass spectrometry. BMC Bioinformatics 13:208. doi:10.1186/1471-2105-13-208
15. Schwahn AB, Wong JW, Downard KM (2009) Subtyping of the influenza virus by high resolution mass spectrometry. Anal Chem 81(9):3500–3506. doi:10.1021/ac900026f
16. Wong JW, Schwahn AB, Downard KM (2010) FluTyper-an algorithm for automated typing and subtyping of the influenza virus from high resolution mass spectral data. BMC Bioinformatics 11:266. doi:10.1186/1471-2105-11-266
17. Frank AM (2009) Predicting intensity ranks of peptide fragment ions. J Proteome Res 8(5):2226–2240. doi:10.1021/pr800677f
18. Han MV, Zmasek CM (2009) PhyloXML: XML for evolutionary biology and comparative genomics. BMC Bioinformatics 10:356. doi:10.1186/1471-2105-10-356
19. Chambers MC, Maclean B, Burke R, Amodei D, Ruderman DL, Neumann S, Gatto L, Fischer B, Pratt B, Egertson J, Hoff K, Kessner D, Tasman N, Shulman N, Frewen B, Baker TA, Brusniak MY, Paulse C, Creasy D, Flashner L, Kani K, Moulding C, Seymour SL, Nuwaysir LM, Lefebvre B, Kuhlmann F, Roark J, Rainer P, Detlev S, Hemenway T, Huhmer A, Langridge J, Connolly B, Chadick T, Holly K, Eckels J, Deutsch EW, Moritz RL, Katz JE, Agus DB, MacCoss M, Tabb DL, Mallick P (2012) A cross-platform toolkit for mass spectrometry and proteomics. Nat Biotechnol 30(10):918–920. doi:10.1038/nbt.2377
20. Perkins DN, Pappin DJ, Creasy DM, Cottrell JS (1999) Probability-based protein identification by searching sequence databases using mass spectrometry data. Electrophoresis 20(18):3551–3567. doi:10.1002/(SICI)1522-2683(19991201)20:18<3551::AID-ELPS3551>3.0.CO;2-2
21. Smith TF, Waterman MS (1981) Identification of common molecular subsequences. J Mol Biol 147(1):195–197
22. UniProt C (2015) UniProt: a hub for protein information. Nucleic Acids Res 43(Database issue):D204–D212. doi:10.1093/nar/gku989
23. Larkin MA, Blackshields G, Brown NP, Chenna R, McGettigan PA, McWilliam H, Valentin F, Wallace IM, Wilm A, Lopez R, Thompson JD, Gibson TJ, Higgins DG (2007) Clustal W and Clustal X version 2.0. Bioinformatics 23(21):2947–2948. doi:10.1093/bioinformatics/btm404
24. Sievers F, Wilm A, Dineen D, Gibson TJ, Karplus K, Li W, Lopez R, McWilliam H, Remmert M, Soding J, Thompson JD, Higgins DG (2011) Fast, scalable generation of high-quality protein multiple sequence alignments using Clustal Omega. Mol Syst Biol 7:539. doi:10.1038/msb.2011.75
25. Edgar RC (2004) MUSCLE: multiple sequence alignment with high accuracy and high throughput. Nucleic Acids Res 32(5):1792–1797. doi:10.1093/nar/gkh340
26. Bandeira N, Clauser KR, Pevzner PA (2007) Shotgun protein sequencing: assembly of peptide tandem mass spectra from mixtures of modified proteins. Mol Cell Proteomics 6(7):1123–1134. doi:10.1074/mcp.M700001-MCP200
27. Henikoff S, Henikoff JG (1992) Amino acid substitution matrices from protein blocks. Proc Natl Acad Sci U S A 89(22):10915–10919
28. Hall BG (2005) Comparison of the accuracies of several phylogenetic methods using protein and DNA sequences. Mol Biol Evol 22(3):792–802. doi:10.1093/molbev/msi066
29. Yang Z, Rannala B (2012) Molecular phylogenetics: principles and practice. Nat Rev Genet 13(5):303–314. doi:10.1038/nrg3186
30. Hoopmann MR, Finney GL, MacCoss MJ (2007) High-speed data reduction, feature detection, and MS/MS spectrum quality assessment of shotgun proteomics data sets using high-resolution mass spectrometry. Anal Chem 79(15):5620–5632. doi:10.1021/ac0700833
31. Brown TA (2002) Molecular phylogenetics. In: Genomes. Wiley-Liss, Oxford
32. Gupta RS (1998) Protein phylogenies and signature sequences: a reappraisal of evolutionary relationships among archaebacteria, eubacteria,

and eukaryotes. Microbiol Mol Biol Rev 62(4):1435–1491

33. Thompson JD, Linard B, Lecompte O, Poch O (2011) A comprehensive benchmark study of multiple sequence alignment methods: current challenges and future perspectives. PLoS One 6(3), e18093. doi:10.1371/journal.pone.0018093

Chapter 12

Bioinformatics Methods to Deduce Biological Interpretation from Proteomics Data

Krishna Patel, Manika Singh, and Harsha Gowda

Abstract

High-throughput proteomics studies generate large amounts of data. Biological interpretation of these large scale datasets is often challenging. Over the years, several computational tools have been developed to facilitate meaningful interpretation of large-scale proteomics data. In this chapter, we describe various analyses that can be performed and bioinformatics tools and resources that enable users to do the analyses. Many Web-based and stand-alone tools are relatively user-friendly and can be used by most biologists without significant assistance.

Key words Gene ontology, FunRich, Reactome, NetPath, Phosphoproteome, Pathways, Enrichment, Post-translational modifications

1 Introduction

High-throughput proteomics studies result in identification and quantitation of thousands of proteins in a biological specimen. These studies are often carried out to determine dynamic changes in proteins including differential expression pattern between biological conditions, activation of specific signaling pathways and in protein complexes. To achieve these, mass spectrometry based methods are often employed to measure relative abundance of proteins or post-translational modifications including phosphorylation, acetylation, glycosylation, and ubiquitination. Although such large-scale studies generate enormous amount of data, they pose significant challenge for biologists for biological interpretation.

Several commercial and open source tools have been developed over the years to facilitate biological interpretation of proteomics data. These tools allow biologists to disentangle complexity in large datasets and identify meaningful patterns. Most biological processes are not driven by a single protein but many proteins acting in concert. If any two biological conditions or cell phenotypes were compared using quantitative proteomics, one could expect a

Shivakumar Keerthikumar and Suresh Mathivanan (eds.), *Proteome Bioinformatics*, Methods in Molecular Biology, vol. 1549, DOI 10.1007/978-1-4939-6740-7_12, © Springer Science+Business Media LLC 2017

set of proteins that regulate these two distinct cell phenotypes or biological conditions to be differentially expressed. Tools that are developed to carry out gene set enrichment or overrepresentation analysis enable identification of such patterns from large scale datasets. Such enrichment analysis can also facilitate functional annotation of orphan molecules based on their association with other well-characterized molecules. Here, we describe several tools that can be used for such analysis in mammalian system, particularly those that have well-annotated data including human.

2 Materials

Several commercial as well as open source tools are available for carrying out bioinformatics analysis of high throughput datasets. For each type of analysis, we are providing list of tools that can be used in relevant sections of the chapter. A step-by-step instruction is also provided for one tool in each section. General outline of the workflow and different kinds of analysis that can be carried out is provided in Fig. 1.

3 Methods

3.1 Gene Ontology Enrichment Analysis

Gene ontology (GO) consortium has developed controlled vocabulary to represent biological functions, processes, and cellular localization information [1]. The terms are linked to corresponding genes based on our understanding of gene function and localization. This data is extensively used to carry out GO enrichment analysis that provides insights into biological functions/processes enriched in a large scale proteomics dataset. There are several tools that have been developed to carry out enrichment analysis providing gene/protein list as an input. FunRich [2] is a user friendly stand-alone tool for GO enrichment analysis. The tool allows users to upload or paste gene symbols, gene ID, Uniprot ID, and RefSeq protein ID as input for the analysis. Results of the enrichment analysis are produced in various graphical formats such as bar graph, pie chart, Venn diagram, heat map, and doughnut chart. Multiple gene sets can be uploaded for comparative analysis of GO enrichment and pathway enrichment analysis. The tool provides various graphical representation options for visualizing comparative results.

One of the widely used Web-based tools is DAVID (Database for Annotation, Visualization, and Integrated Discovery (https://david.ncifcrf.gov/) [3]. It provides a comprehensive set of functional annotation tools which can not only identify enriched biological themes, particularly GO terms, but also discover functionally related enriched gene groups based on popular pathway databases including KEGG [4] and BioCarta [5]. Here we describe a step-by-step guide for GO enrichment using DAVID.

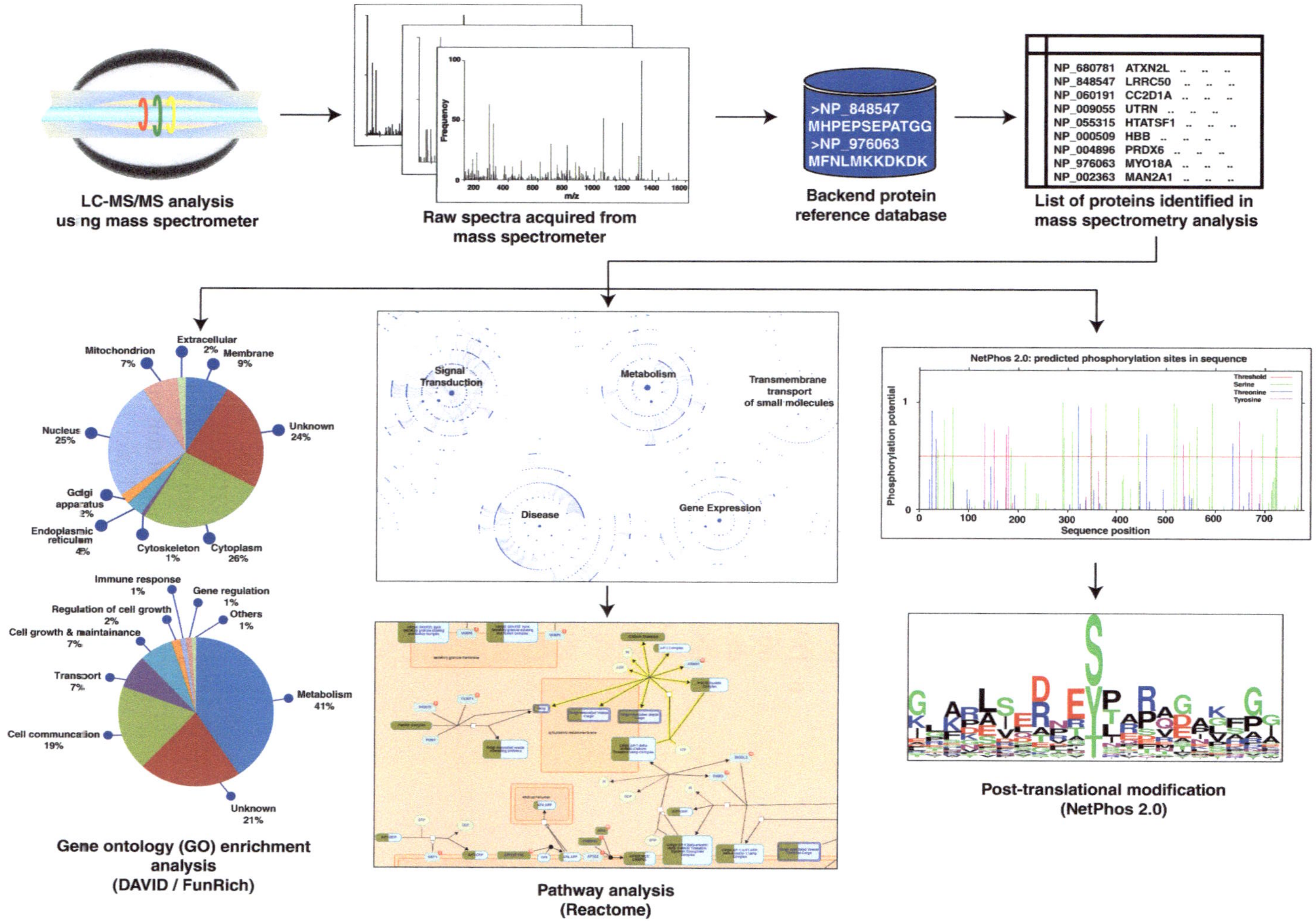

Fig. 1 A general framework and outline of various bioinformatics analyses approaches that can be used for high-throughput proteomic data

There are two major DAVID tools that could be used for functional annotation/classification of gene lists—Functional Annotation and Gene Functional Classification. The tools can be accessed by clicking the links on top left corner of the home page.

1. To begin the analysis, click on "Functional annotation".
2. The resulting Web page shows three tabs—Upload, List, and Background.
3. In the "Upload" tab, either paste gene list into the box or browse and upload the list where there is a single column with each row representing a single gene (*see* **Note 1**).
4. The 'list' tab in DAVID allows users to limit gene annotations to one or more species. The default parameter chooses *Homo sapiens.*
5. For enrichment analysis, user has to choose a background using 'Background' tab. Default background in DAVID is *Homo sapiens* whole genome background. The user can choose to use a custom background.
6. DAVID recognizes gene lists with various identifiers including official gene symbols and accession numbers. For proteomics datasets, it is best to use official gene symbols in gene lists and choose that as an identifier in **step 2** in 'Upload' tab.
7. In **step 3**, choose if the list you uploaded should be used as 'Gene List' or 'Background'. For data from human samples, choose your input as 'Gene List' as *Homo sapiens* whole genome background is used as a default.
8. Click 'Submit List' button. The results provided by DAVID include 'Functional Annotation Clustering', 'Functional Annotation Chart' and 'Functional Annotation Table'. These results provide a quick glance of major biological functions enriched in the gene list.
9. For GO enrichment analysis, click on Gene_Ontology and select GOTERM_BP_ALL for biological process, GOTERM_CC_ALL for subcellular localization, and GOTERM_MF_ALL for molecular function as background for the GO enrichment analysis. Click on "Functional annotation clustering" and DAVID will generate clusters of terms with similar biological meaning based on shared/similar gene members. The significance of this enrichment is also calculated based on modified Fisher Exact *P*-value.
10. Top panel of the result window is parameter panel which user can modify according to need and rerun the process without submitting input again. It is recommended to select higher stringency for small, concise and meaningful clusters rather than broader and vague cluster of proteins. Default setting is medium stringency however user can modify this option based on the analysis.

Higher enrichment score indicates that annotation term members are overrepresented in uploaded input.

11. Result table displays annotation categories, enriched functional annotation, enrichment scores of each cluster, number of genes contributing to clustering of similar GO terms, and modified Fisher Exact *P*-value.
12. To analyze the most enriched clusters, user can sieve out clusters with maximum enrichment score and lesser *P*-value for biological process, molecular function and subcellular localization (*see* **Note 2**).
13. A link to 'G' on top of each cluster could be used to extract defined set of proteins contributing to enrichment of the given cluster and matrix icon draws heat map for the small cluster and provides the GO term count matrix for each protein which can be further used for plotting graphs.
14. User can also employ pathway and functional domain enrichment analysis using DAVID by selecting "Pathway", "Functional categories" and "Protein domains" as backend reference database for functional annotation. However, a user-friendly graphical user interface for pathways analysis study is deployed by Web-resource Reactome which is explained in detail below. Table 1 enlists other widely used open source gene set enrichment analysis tools.

3.2 Pathway Analysis

Proteins regulate most cellular processes. Several proteins work in concert to regulate these processes and are often grouped into specific pathways in which they carry out their functions. Over the years, pathways and processes that are regulated by specific proteins have been systematically annotated. Based on protein expression data, it is possible to arrive at pathways and processes that are active in a biological sample. In addition to expression, some of the most widely studied signaling pathway mechanisms include dynamic interplay of kinases and phosphatases that results in addition or removal of phosphorylation on proteins. Differential protein expression data or phosphoproteomics data can be utilized to carry out pathway enrichment analysis. If expression or phosphorylation levels of certain proteins are changing in a biological sample as compared to their pattern in an appropriate control, it is possible to predict potential pathways that are differentially regulated. Reactome [14] is manually curated open access Web-based resource of biological pathways which allows users to browse, search and map proteins onto pathways. It also provides list of interactors acquired from IntAct [15] molecular interaction database with nodes of pathways.

Here we describe Reactome, a Web-based tool that can be used for pathway analysis.

Table 1
List of tools that can be used for gene ontology and gene set enrichment analysis

Name	Description	Link	Reference
GSEA	Gene set enrichment analysis (GSEA) is an expression analytics tool. It compares gene set enrichment between conditions and provides enriched set of genes with their statistical significance scores to interpret biological data	Stand-alone http://www.broadinstitute.org/gsea/	[6]
FunRich	FunRich is a downloadable tool for pathways and GO enrichment analysis of genes and proteins. It can process genes/proteins irrespective of source of the sample as user can load customized database along with default available background database	Stand-alone http://funrich.org/	[2]
GoMiner	GoMiner leverages Gene Ontology by providing a framework to visualize and integrate "omics" data. It makes cluster of genes and their expression profiles which can be analyzed for their biological significance. Each gene is linked to BioCarta, Entez Genome, NCBI structures, Pubmed and MedMiner for greater clarity	Stand-alone, Web http://discover.nci.nih.gov/gominer	[7]
GOstat	GOstat tool uses GO terms database to find statistically over represented genes from the data set. The results list out significant set of genes for biological interpretation	Web http://gostat.wehi.edu.au	[8]
GOToolBox	GOToolBox is used for functional annotation of genes. GOtoolBox is a perl based program which can be automated in any gene expression analysis pipeline. GOToolBox also has GO-Diet and PRODISTIN framework which can be used to study protein–protein interactions	Web http://genome.crg.es/GOToolBox/	[9]

(continued)

Table 1
(continued)

Name	Description	Link	Reference
GeneMerge	GeneMerge enables over-representation analysis of gene attributes in a given set of genes as compared to genome background	Stand-alone, Web http://www.genemerge.net/	[10]
GO:TermFinder	GO:TermFinder is a tool that helps to find significant GO terms shared among a list of genes. It has GO:TermFinder libraries that enables visualization of results	Stand-alone http://search.cpan.org/dist/GO-TermFinder/	[11]
agriGO	agriGO is a specialized data analytics tool for the agricultural community. The database has 38 agricultural species comprising of 274 data types	Web http://bioinfo.cau.edu.cn/agriGO/	[12]
FatiGO	FatiGO helps to find significant over-representation of functional annotations in one gene set compared to the other	Web http://babelomics.bioinfo.cipf.es	[13]

1. Reactome (http://www.reactome.org/) allows mapping the list of proteins on pathways and carry out enrichment analysis to determine if the input data contains overrepresentation of proteins involved in certain pathways (*see* **Note 3**).
2. Click on "Analyze Data". It is a three-step process that begins with pasting the protein list with appropriate header on the Web page. The tool also takes accession numbers and other identifiers as an input. In the next step, it allows projection of data on to human annotation if it comes from a different species and also to include interactors from IntAct Molecular Interaction database. After making appropriate selection, click on analyze.
3. The resulting page is divided into four panels. 'Hierarchy panel' on the left part of the Web page lists enriched pathways with corresponding FDR, 'Viewport' panel shows graphical representation of an overview of these pathways with various options to navigate, top panel provides configuration options and a bottom panel provides details of objects selected in the pathway diagram. A detailed manual to understand and navigate this pathway analysis tool can be found at http://wiki.reactome.org/index.php/Usersguide.

There are various commercial tools such as QIAGEN Ingenuity Pathway Analysis (IPA) and Agilent Genomics Genespring for functional and pathway enrichment analysis. Table 2 lists some of the widely used pathway resources and network analysis tools.

3.3 Post-translational Modification Analysis

Post-translational modifications (PTM) play an important role in regulating various cellular processes. One of the most widely studied PTM is phosphorylation. It acts as a switch for activation and deactivation of specific proteins and associated signaling pathways. This modification serves as a rapid and reversible means to modulate protein activity and transduce signals. Advent of mass spectrometry has revolutionized our ability to map PTMs. These studies have provided a comprehensive view of proteins that undergo modifications along with specific sites. Based on our understanding of enzyme–substrate relationships and specific motifs that are targeted for post-translational modifications, a number of computational tools have been developed to predict PTMs. These tools can be utilized to evaluate the validity of identified sites in large scale studies (based on known sites in the database) or predict potential modifications.

Human Protein Reference Database (HPRD) [21] is a repository of manually curated PTM sites. Phospho.ELM [22] is a resource of experimentally validated phosphorylation sites that are manually curated from the literature. The RESID [23] database provides PTM information with literature citation, protein feature table, molecular models, structure diagrams and Gene Ontology cross reference. PhosphoSitePlus [24] is a comprehensive repository of curated phosphosites containing reference and orthologous residues in other species. O-GLYCBASE [25] is a resource containing experimentally verified O-linked glycosylation sites. Unimod [26] is a comprehensive public domain database of protein modifications for mass spectrometry application.

Most extensively studied PTM is phosphorylation. Protein kinases add phosphate moieties to Tyr, Ser, or Thr residues. Mass spectrometry is being extensively used to investigate protein phosphorylation in a high-throughput manner. Phosphorylation either increases or decreases the activity of target protein. Overlaying phosphoproteomic data on curated pathways can provide insights into activation or deactivation of a particular signaling pathway. PhosphositePlus [24] and PHOSIDA [27] are comprehensive repositories of curated phosphosites containing reference and orthologous residues in other species. Protein sequences can be analyzed using various prediction tools for identifying phosphosites such as KinasePhos 2.0 [28], NetPhos 2.0 [29], and DISPHOS 1.3 [30].

Several computational approaches have been developed to predict acetylation sites. NetAcet [31] is a neural network based N-terminal acetylation site prediction tool, N-Ace [32] predicts acetylation sites based on physicochemical properties of protein with accessible surface area, PSKAcePred [33] is an approach that uses

Table 2
List of pathway resources and network analysis tools

Name	Description	Link	Reference
NetPath	NetPath is a manually curated resource of signal transduction pathways. Pathway data can be browsed, visualized or downloaded in PSI-MI, BioPAX and SBML formats. These standard formats enable visualization using external tools like Cytoscape	Web www.netpath.org	[16]
PANTHER	Protein ANalysis THrough Evolutionary Relationships (PANTHER) is an analysis framework with multiple tools for evolutionary and functional classification of proteins. Panther pathway resource allows visualization of protein expression data in the context of pathway diagrams	Web http://www.pantherdb.org/pathway	[17]
KEGG	Kyoto encyclopedia of genes and genomes (KEGG) is an integrated database resource. Pathway maps and annotation in KEGG is widely used for pathway enrichment analysis	Web http://www.genome.jp/kegg/	[4]
STRING	Search Tool for the Retrieval of Interacting Genes/Proteins (STRING) is a database of protein–protein interactions	Web http://string-db.org/	[18]
FunRich	FunRich is a downloadable tool for pathways and GO enrichment analysis of genes and proteins. It can process genes/proteins irrespective of source of the sample as user can load customized database along with default available background database	Stand-alone http://funrich.org/	[2]
MINT	MINT: Molecular INTeraction is a curated molecular interaction database	Web, stand-alone http://mint.bio.uniroma2.it/mint/Welcome.do	[19]
NetworKIN	NetworKIN database provides interface to analyze cellular phosphorylation networks. It allows users to query precomputed kinase–substrate relations or obtain predictions on novel phosphoproteins	Web, stand-alone http://networkin.info	[20]

evolutionary similarity along with physicochemical properties to predict lysine acetylation sites and Species Specific Prediction of Lysine Acetylation (SSPKA) [34] is a computational framework that incorporates predicted secondary structure information, and combines functional features and sequence feature to predict species-specific acetylation sites across six different species—*H. sapiens*, *R. norvegicus*, *M. musculus*, *E. coli*, *S. typhimurium and S. cerevisiae.*

Ubiquitination is one of the most difficult PTMs to be identified due to its low abundance, size, and dynamic regulation. Due to larger size of ubiquitin compared to other PTMs, it is difficult to capture by mass spectrometry. However, several ubiquitination sites have been mapped in the last few years based on diglycine-modified lysine tag can be identified by mass spectrometry. Several tools including UbPred [35], UbiPred [36], E3Miner [37], hCK-SAAP_UbSite [38], and iUbiq-Lys [39] have been developed over the years for prediction of ubiquitination sites. hUbiquitome [40] is a comprehensive repository of experimentally verified human ubiquitination enzymes and substrates.

Small ubiquitin-like modifier (SUMO) attaches to various target proteins and modulates cellular processes such as DNA replication, transcription, cell division, nuclear trafficking, and DNA damage response. SUMOylation affects half-life, localization of targets or binding partners and is a crucial mechanism that allows cells to adapt to stress stimuli. Identification of SUMO sites has enabled us to identify strong dependency of SUMOylation events on other PTMs [41]. SUMOsp [42] and GPS-SUMO [43] predicts SUMO sites on proteins.

Glycosylation is a common PTM that plays a crucial role in protein folding, cell–cell interaction, antigenicity, transport, and half-life. There are four types of glycosylation: N-linked, O-linked, C-mannosylation, and GPI anchor attachment. EnsembleGly [44] predicts both O- and N-linked glycosylation sites, NetCGlyc [45] predicts C-mannosylation, NetOGlyc [46] predicts O-glycosylation sites, and NetNGlyc [47] predicts N-Glycosylation sites; PredGPI [48] and GPI-SOM [49] predict GPI anchor sites in a protein.

Scansite [50] is a tool to analyze protein sequence for phosphorylation motifs recognized by many kinases and Motif-X [51] allows prediction of various PTM site motifs by identifying over-represented residues in the flanking regions. ProMEX [52] is a database of mass spectra of tryptic peptides from plant proteins and phosphoproteins.

Here we describe PTM analysis using commonly used PTM database Phospho.ELM [22] and phosphorylation PTM site predictor NetPhos 2.0 [29].

1. To identify experimentally validated PTMs of a given protein, browse Phospho.ELM database (http://phospho.elm.eu.org/index.html). Database can be queried using protein name, UniPROT accession, and Ensembl identifier.

2. Result page of Phospho.ELM database consists of table detailing residue, position of residue in proteins, flanking sequence with PTM site, kinase, PubMed reference for each site reported, conservation score, cross-reference to eukaryotic linear motif resource (ELM: http://elm.eu.org/), phospho-peptide binding domain, SMART domains, and cross-reference to PDB link along with other information such as substrate, cross-reference to PHOSIDA [27], PhosphositePlus [24], MINT [19], and GO-Terms [1].
3. Computational prediction of phosphorylation can be done using NetPhos 2.0 server (http://www.cbs.dtu.dk/services/NetPhos/). Users can submit protein sequence in FASTA format and select target residue for phosphorylation (tyrosine, serine, or threonine). By default, all three residues are checked in the analysis. Select checkbox if users wish to generate graphical output.
4. Click on "Submit" to initiate analysis. In a single query, up to 2000 protein sequences can be analyzed by this Web-based tool.
5. Result page will display table detailing submitted protein ID, residue position, PTM site with flanking sequences and score. Three tables are separately generated for serine, threonine, and tyrosine.
6. A graphical result depicts propensity of a residue on a given position as PTM site. Three different color peaks are used for each residue (S,T,Y) on an *X*-*Y* plane where *X*-axis is sequence position and *Y*-axis is phosphorylation potential.

3.4 Visualization Tools

A multitude of tools are available for data integration and visualization of "omics" data-sets (Table 3). Most visualization tools focus on biomolecular interactions and pathways. These tools commonly employ 2D graphs for data representation. The basic efficiency of these tools lies in its compatibility with other tools and databases.

4 Notes

1. It is preferable to use 'Gene Symbol' as unique identifier for genes. DAVID has ID conversion tool that can be used to prepare the lists with uniform identifiers.
2. Enrichment analysis methods often involve statistical tests to determine if input data contains overrepresentation of proteins involved in certain functions, processes, or pathways more than what is expected by chance. This is calculated with respect to the background database used by respective tools. Many tools also provide flexibility for users by providing the option of using custom database as background. Knowledge of statistical

Table 3
List of pathway analysis and visualization tools

Name	Description	Link	Reference
GenMAPP	GenMAPP is a Web-based visualization tool for gene/protein expression profiles. It has MAPPBuilder tool for creating MAPP file (.mapp) which creates graphical pathway representation of genes and MAPPFinder tool to annotate the pathway. Each gene is identified by unique geneID from Genbank. MAPP files can be shared and manipulated by the user	Stand-alone http://www.genmapp.org	[53]
CytoScape	Cytoscape is Java-based stand-alone tool which supports large scale network analysis. Both protein–protein and protein–gene networks can be visualized and edited. The standard file format of Cytoscape is Cytoscape Session File (.cys). Input file in Cytoscape can be delimited text table or excel workbook though it supports all major input formats. The result can be exported in any of the formats like SIF, GML, XGMML, and PSI-MI formats	Stand-alone http://www.cytoscape.org/	[54]
Medusa	Medusa is Java application for visualization of complex pathways. Result from STRING pathway database can be analyzed in Medusa. Medusa is less suited for big datasets	Stand-alone https://sites.google.com/site/medusa3visualization/	[55]
Perseus	Perseus is a statistical analysis visualization tool for proteomics data. It has incorporated multiple statistical methods like *t*-test, clustering, enrichment analysis including normalization of data. It provides various graphs for visualization of data like scatter plot and volcano plot	Stand-alone http://www.biochem.mpg.de/5111810/perseus	[56]

approach employed in such tools would allow user to make relevant selections for different kind of datasets to identify most enriched genes/proteins cluster.

3. Pathway enrichment analysis is done using the pathway database used in the background. Back end pathway database used for analysis will directly influence the outcomes of the pathway analysis. This aspect should be taken into consideration and users should select appropriate pathway annotation resource most suitable for intended pathway analysis.

References

1. Ashburner M, Ball CA, Blake JA, Botstein D, Butler H, Cherry JM, Davis AP, Dolinski K, Dwight SS, Eppig JT, Harris MA, Hill DP, Issel-Tarver L, Kasarskis A, Lewis S, Matese JC, Richardson JE, Ringwald M, Rubin GM, Sherlock G (2000) Gene ontology: tool for the unification of biology. The gene ontology consortium. Nat Genet 25(1):25–29. doi:10.1038/75556
2. Pathan M, Keerthikumar S, Ang CS, Gangoda L, Quek CY, Williamson NA, Mouradov D, Sieber OM, Simpson RJ, Salim A, Bacic A, Hill AF, Stroud DA, Ryan MT, Agbinya JI, Mariadason JM, Burgess AW, Mathivanan S (2015) FunRich: an open access standalone functional enrichment and interaction network analysis tool. Proteomics 15(15):2597–2601. doi:10.1002/pmic.201400515
3. Dennis G Jr, Sherman BT, Hosack DA, Yang J, Gao W, Lane HC, Lempicki RA (2003) DAVID: database for annotation, visualization, and integrated discovery. Genome Biol 4(5):P3
4. Kanehisa M, Goto S (2000) KEGG: Kyoto encyclopedia of genes and genomes. Nucleic Acids Res 28(1):27–30
5. Nishimura D (2004) BioCarta. Biotech Software Internet Report 2:117–120. doi: 10.1089/152791601750294344
6. Subramanian A, Tamayo P, Mootha VK, Mukherjee S, Ebert BL, Gillette MA, Paulovich A, Pomeroy SL, Golub TR, Lander ES, Mesirov JP (2005) Gene set enrichment analysis: a knowledge-based approach for interpreting genome-wide expression profiles. Proc Natl Acad Sci U S A 102(43):15545–15550. doi:10.1073/pnas.0506580102
7. Zeeberg BR, Feng W, Wang G, Wang MD, Fojo AT, Sunshine M, Narasimhan S, Kane DW, Reinhold WC, Lababidi S, Bussey KJ, Riss J, Barrett JC, Weinstein JN (2003) GoMiner: a resource for biological interpretation of genomic and proteomic data. Genome Biol 4(4):R28
8. Beissbarth T, Speed TP (2004) GOstat: find statistically overrepresented gene ontologies within a group of genes. Bioinformatics 20(9):1464–1465. doi:10.1093/bioinformatics/bth088
9. Martin D, Brun C, Remy E, Mouren P, Thieffry D, Jacq B (2004) GOToolBox: functional analysis of gene datasets based on Gene Ontology. Genome Biol 5(12):R101. doi:10.1186/gb-2004-5-12-r101
10. Castillo-Davis CI, Hartl DL (2003) GeneMerge—post-genomic analysis, data mining, and hypothesis testing. Bioinformatics 19(7):891–892
11. Boyle EI, Weng S, Gollub J, Jin H, Botstein D, Cherry JM, Sherlock G (2004) GO::TermFinder—open source software for accessing gene ontology information and finding significantly enriched gene ontology terms associated with a list of genes. Bioinformatics 20(18):3710–3715. doi:10.1093/bioinformatics/bth456
12. Du Z, Zhou X, Ling Y, Zhang Z, Su Z (2010) agriGO: a GO analysis toolkit for the agricultural community. Nucleic Acids Res 38(Web Server Issue):64–70. doi:10.1093/nar/gkq310
13. Al-Shahrour F, Minguez P, Tarraga J, Medina I, Alloza E, Montaner D, Dopazo J (2007) FatiGO +: a functional profiling tool for genomic data. Integration of functional annotation, regulatory motifs and interaction data with microarray experiments. Nucleic Acids Res 35(Web Server Issue):91–96. doi:10.1093/nar/gkm260
14. Joshi-Tope G, Gillespie M, Vastrik I, D'Eustachio P, Schmidt E, de Bono B, Jassal B, Gopinath GR, Wu GR, Matthews L, Lewis S, Birney E, Stein L (2005) Reactome: a knowledgebase of biological pathways. Nucleic Acids Res 33(Database issue):D428–D432. doi:10.1093/nar/gki072
15. Hermjakob H, Montecchi-Palazzi L, Lewington C, Mudali S, Kerrien S, Orchard S, Vingron M, Roechert B, Roepstorff P, Valencia A, Margalit H, Armstrong J, Bairoch A, Cesareni G, Sherman D, Apweiler R (2004) IntAct: an open source molecular interaction database. Nucleic Acids Res 32(Database issue):D452–D455. doi:10.1093/nar/gkh052
16. Kandasamy K, Mohan SS, Raju R, Keerthikumar S, Kumar GS, Venugopal AK, Telikicherla D, Navarro JD, Mathivanan S, Pecquet C, Gollapudi SK, Tattikota SG, Mohan S, Padhukasahasram H, Subbannayya Y, Goel R, Jacob HK, Zhong J, Sekhar R, Nanjappa V, Balakrishnan L, Subbaiah R, Ramachandra YL, Rahiman BA, Prasad TS, Lin JX, Houtman JC, Desiderio S, Renauld JC, Constantinescu SN, Ohara O, Hirano T, Kubo M, Singh S, Khatri P, Draghici S, Bader GD, Sander C, Leonard WJ, Pandey A (2010) NetPath: a public resource of curated signal transduction pathways. Genome Biol 11(1):R3. doi:10.1186/gb-2010-11-1-r3

17. Mi H, Poudel S, Muruganujan A, Casagrande JT, Thomas PD (2016) PANTHER version 10: expanded protein families and functions, and analysis tools. Nucleic Acids Res 44(D1):D336–D342. doi:10.1093/nar/gkv1194
18. von Mering C, Huynen M, Jaeggi D, Schmidt S, Bork P, Snel B (2003) STRING: a database of predicted functional associations between proteins. Nucleic Acids Res 31(1):258–261
19. Zanzoni A, Montecchi-Palazzi L, Quondam M, Ausiello G, Helmer-Citterich M, Cesareni G (2002) MINT: a molecular INTeraction database. FEBS Lett 513(1):135–140
20. Linding R, Jensen LJ, Pasculescu A, Olhovsky M, Colwill K, Bork P, Yaffe MB, Pawson T (2008) NetworKIN: a resource for exploring cellular phosphorylation networks. Nucleic Acids Res 36(Database issue):D695–D699. doi:10.1093/nar/gkm902
21. Peri S, Navarro JD, Kristiansen TZ, Amanchy R, Surendranath V, Muthusamy B, Gandhi TK, Chandrika KN, Deshpande N, Suresh S, Rashmi BP, Shanker K, Padma N, Niranjan V, Harsha HC, Talreja N, Vrushabendra BM, Ramya MA, Yatish AJ, Joy M, Shivashankar HN, Kavitha MP, Menezes M, Choudhury DR, Ghosh N, Saravana R, Chandran S, Mohan S, Jonnalagadda CK, Prasad CK, Kumar-Sinha C, Deshpande KS, Pandey A (2004) Human protein reference database as a discovery resource for proteomics. Nucleic Acids Res 32(Database issue):D497–D501. doi:10.1093/nar/gkh070
22. Diella F, Cameron S, Gemund C, Linding R, Via A, Kuster B, Sicheritz-Ponten T, Blom N, Gibson TJ (2004) Phospho.ELM: a database of experimentally verified phosphorylation sites in eukaryotic proteins. BMC Bioinformatics 5:79. doi:10.1186/1471-2105-5-79
23. Garavelli JS (2004) The RESID database of protein modifications as a resource and annotation tool. Proteomics 4(6):1527–1533. doi:10.1002/pmic.200300777
24. Hornbeck PV, Zhang B, Murray B, Kornhauser JM, Latham V, Skrzypek E (2015) PhosphoSitePlus, 2014: mutations, PTMs and recalibrations. Nucleic Acids Res 43(Database issue):D512–D520. doi:10.1093/nar/gku1267
25. Gupta R, Birch H, Rapacki K, Brunak S, Hansen JE (1999) O-GLYCBASE version 4.0: a revised database of O-glycosylated proteins. Nucleic Acids Res 27(1):370–372
26. Creasy DM, Cottrell JS (2004) Unimod: protein modifications for mass spectrometry. Proteomics 4(6):1534–1536. doi:10.1002/pmic.200300744
27. Gnad F, Ren S, Cox J, Olsen JV, Macek B, Oroshi M, Mann M (2007) PHOSIDA (phosphorylation site database): management, structural and evolutionary investigation, and prediction of phosphosites. Genome Biol 8(11):R250. doi:10.1186/gb-2007-8-11-r250
28. Huang HD, Lee TY, Tzeng SW, Horng JT (2005) KinasePhos: a web tool for identifying protein kinase-specific phosphorylation sites. Nucleic Acids Res 33(Web Server Issue):226–229. doi:10.1093/nar/gki471
29. Blom N, Gammeltoft S, Brunak S (1999) Sequence and structure-based prediction of eukaryotic protein phosphorylation sites. J Mol Biol 294(5):1351–1362. doi:10.1006/jmbi.1999.3310
30. Iakoucheva LM, Radivojac P, Brown CJ, O'Connor TR, Sikes JG, Obradovic Z, Dunker AK (2004) The importance of intrinsic disorder for protein phosphorylation. Nucleic Acids Res 32(3):1037–1049. doi:10.1093/nar/gkh253
31. Kiemer L, Bendtsen JD, Blom N (2005) NetAcet: prediction of N-terminal acetylation sites. Bioinformatics 21(7):1269–1270. doi:10.1093/bioinformatics/bti130
32. Lee TY, Hsu JB, Lin FM, Chang WC, Hsu PC, Huang HD (2010) N-Ace: using solvent accessibility and physicochemical properties to identify protein N-acetylation sites. J Comput Chem 31(15):2759–2771. doi:10.1002/jcc.21569
33. Suo SB, Qiu JD, Shi SP, Sun XY, Huang SY, Chen X, Liang RP (2012) Position-specific analysis and prediction for protein lysine acetylation based on multiple features. PLoS One 7(11), e49108. doi:10.1371/journal.pone.0049108
34. Li Y, Wang M, Wang H, Tan H, Zhang Z, Webb GI, Song J (2014) Accurate in silico identification of species-specific acetylation sites by integrating protein sequence-derived and functional features. Sci Rep 4:5765. doi:10.1038/srep05765
35. Radivojac P, Vacic V, Haynes C, Cocklin RR, Mohan A, Heyen JW, Goebl MG, Iakoucheva LM (2010) Identification, analysis, and prediction of protein ubiquitination sites. Proteins 78(2):365–380. doi:10.1002/prot.22555
36. Tung CW, Ho SY (2008) Computational identification of ubiquitylation sites from protein sequences. BMC Bioinformatics 9:310. doi: 10.1186/1471-2105-9-310
37. Lee H, Yi GS, Park JC (2008) E3Miner: a text mining tool for ubiquitin-protein ligases. Nucleic Acids Res 36(Web Server Issue):416–422. doi:10.1093/nar/gkn286
38. Chen Z, Zhou Y, Song J, Zhang Z (2013) hCKSAAP_UbSite: improved prediction of human ubiquitination sites by exploiting amino acid pattern and properties. Biochim Biophys Acta 1834(8):1461–1467. doi:10.1016/j.bbapap.2013.04.006

39. Qiu WR, Xiao X, Lin WZ, Chou KC (2015) iUbiq-Lys: prediction of lysine ubiquitination sites in proteins by extracting sequence evolution information via a gray system model. J Biomol Struct Dyn 33(8):1731–1742. doi:10.1080/07391102.2014.968875
40. Du Y, Xu N, Lu M, Li T (2011) hUbiquitome: a database of experimentally verified ubiquitination cascades in humans. Database (Oxford) 2011:bar055. doi:10.1093/database/bar055
41. Eifler K, Vertegaal AC (2015) Mapping the SUMOylated landscape. FEBS J 282(19):3669–3680. doi:10.1111/febs.13378
42. Xue Y, Zhou F, Fu C, Xu Y, Yao X (2006) SUMOsp: a web server for sumoylation site prediction. Nucleic Acids Res 34(Web Server Issue):254–257. doi:10.1093/nar/gkl207
43. Zhao Q, Xie Y, Zheng Y, Jiang S, Liu W, Mu W, Liu Z, Zhao Y, Xue Y, Ren J (2014) GPS-SUMO: a tool for the prediction of sumoylation sites and SUMO-interaction motifs. Nucleic Acids Res 42(Web Server Issue):325–330. doi:10.1093/nar/gku383
44. Caragea C, Sinapov J, Silvescu A, Dobbs D, Honavar V (2007) Glycosylation site prediction using ensembles of Support Vector Machine classifiers. BMC Bioinformatics 8:438. doi:10.1186/1471-2105-8-438
45. Julenius K (2007) NetCGlyc 1.0: prediction of mammalian C-mannosylation sites. Glycobiology 17(8):868–876. doi:10.1093/glycob/cwm050
46. Hansen JE, Lund O, Tolstrup N, Gooley AA, Williams KL, Brunak S (1998) NetOglyc: prediction of mucin type O-glycosylation sites based on sequence context and surface accessibility. Glycoconj J 15(2):115–130
47. Gupta R, Jung E, Brunak S (2004) NetNGlyc 1.0 Server. Center for biological sequence analysis, technical university of Denmark (http://wwwcbsdtudk/services/NetNGlyc)
48. Pierleoni A, Martelli PL, Casadio R (2008) PredGPI: a GPI-anchor predictor. BMC Bioinformatics 9:392. doi:10.1186/1471-2105-9-392
49. Fankhauser N, Maser P (2005) Identification of GPI anchor attachment signals by a Kohonen self-organizing map. Bioinformatics 21(9):1846–1852. doi:10.1093/bioinformatics/bti299
50. Obenauer JC, Cantley LC, Yaffe MB (2003) Scansite 2.0: proteome-wide prediction of cell signaling interactions using short sequence motifs. Nucleic Acids Res 31(13):3635–3641
51. Chou MF, Schwartz D (2011) Biological sequence motif discovery using motif-x. Curr Protoc Bioinformatics 13:15–24. doi:10.1002/0471250953.bi1315s35
52. Hummel J, Niemann M, Wienkoop S, Schulze W, Steinhauser D, Selbig J, Walther D, Weckwerth W (2007) ProMEX: a mass spectral reference database for proteins and protein phosphorylation sites. BMC Bioinformatics 8:216. doi:10.1186/1471-2105-8-216
53. Dahlquist KD, Salomonis N, Vranizan K, Lawlor SC, Conklin BR (2002) GenMAPP, a new tool for viewing and analyzing microarray data on biological pathways. Nat Genet 31(1):19–20. doi:10.1038/ng0502-19
54. Shannon P, Markiel A, Ozier O, Baliga NS, Wang JT, Ramage D, Amin N, Schwikowski B, Ideker T (2003) Cytoscape: a software environment for integrated models of biomolecular interaction networks. Genome Res 13(11):2498–2504. doi:10.1101/gr.1239303
55. Hooper SD, Bork P (2005) Medusa: a simple tool for interaction graph analysis. Bioinformatics 21(24):4432–4433. doi:10.1093/bioinformatics/bti696
56. Tyanova S, Temu T, Sinitcyn P, Carlson A, Hein M, Geiger T, Mann M and Cox J (2016) The Perseus computational platform for comprehensive analysis of (prote)omics data. Nature Methods 3(9):731–740. doi:10.1038/nmeth.3901

Chapter 13

A Systematic Bioinformatics Approach to Identify High Quality Mass Spectrometry Data and Functionally Annotate Proteins and Proteomes

Mohammad Tawhidul Islam, Abidali Mohamedali, Seong Beom Ahn, Ishmam Nawar, Mark S. Baker, and Shoba Ranganathan

Abstract

In the past decade, proteomics and mass spectrometry have taken tremendous strides forward, particularly in the life sciences, spurred on by rapid advances in technology resulting in generation and conglomeration of vast amounts of data. Though this has led to tremendous advancements in biology, the interpretation of the data poses serious challenges for many practitioners due to the immense size and complexity of the data. Furthermore, the lack of annotation means that a potential gold mine of relevant biological information may be hiding within this data. We present here a simple and intuitive workflow for the research community to investigate and mine this data, not only to extract relevant data but also to segregate usable, quality data to develop hypotheses for investigation and validation. We apply an MS evidence workflow for verifying peptides of proteins from one's own data as well as publicly available databases. We then integrate a suite of freely available bioinformatics analysis and annotation software tools to identify homologues and map putative functional signatures, gene ontology and biochemical pathways. We also provide an example of the functional annotation of missing proteins in human chromosome 7 data from the NeXtProt database, where no evidence is available at the proteomic, antibody, or structural levels. We give examples of protocols, tools and detailed flowcharts that can be extended or tailored to interpret and annotate the proteome of any novel organism.

Key words MS validation, MS evidence, Functional annotation, Missing proteins

1 Introduction

The advent of high-throughput proteomic and genomic analyses methods ranging from RNASeq to high-end mass spectrometry (MS) has resulted in scientists now being able to generate vast amounts of data from single experiments. The challenge over the past decade has been not only in generating relevant data, but more importantly in manipulating, storing, and interpreting this data to answer pertinent biological questions [1]. Biostatistics and

Shivakumar Keerthikumar and Suresh Mathivanan (eds.), *Proteome Bioinformatics*, Methods in Molecular Biology, vol. 1549, DOI 10.1007/978-1-4939-6740-7_13, © Springer Science+Business Media LLC 2017

bioinformatics tools have become indispensable in extracting and analyzing these data sets [2] but a bewildering array of such tools are available, often each with its own intricacies, interpretation challenges, and shortfalls [3]. However, before statistical analyses and interpreting the results from proteomics data, it is imperative that a protein or gene is correctly identified and, more importantly, annotated. Proteomics data and search engine software in most cases rely on carefully curated and computationally annotated protein databases for protein identification, the results of which are then be further or concurrently analyzed for biological relevance (location, biological/molecular/biochemical processes) with a statistical perspective [4–6]. These annotated proteins are then used as a basis for identifying proteins from proteomics experiments where computationally determined theoretical peptide masses are compared to mass data obtained from often high-accuracy instruments. These mass-based search software programs then report confidence values against an identified protein. Numerous software solutions exist to identify and interpret MS data and are in use as standard operating protocols in laboratories across the globe [7, 8]. Users are then encouraged/required to deposit data into public MS data repositories, which then report protein identifications with details of results from these tools. The primary limitations to correct protein identification include but are not limited to user selected error thresholds, search databases and spectral quality thresholds [9]. Moreover, practitioners are not versed in exactly how the enormous MS data resources can be assessed and used to validate their own MS data.

Another serious obstacle to proteomic studies is often in the analysis of MS data from unannotated (novel) proteomes or unannotated database entries. At the core of most automated annotation technologies is a sequence homology search against protein databases of known function. Most annotation approaches use a single pass searching methodology against one or several databases, which may lead to the generation of unannotated results, putative proteins and translated coding regions without much metadata [10–12]. To overcome these limitations, we have previously reported a novel methodology for sequentially searching multiple databases to allow a more robust drill down [10]. We proposed a sequential search approach, where unannotated proteins are matched against reviewed proteins with experimental evidence, to identify homologues from which putative function assignment can be made [12]. Those proteins for which no database matches were found, are then matched against reviewed proteins. Following this step, the sequences without matches are checked to determine whether they are specific to that organism, due to speciation, followed by searching against proteins with three-dimensional structure, for remote homologue identification. Independently, we have adopted an annotation strategy that assigns putative functional

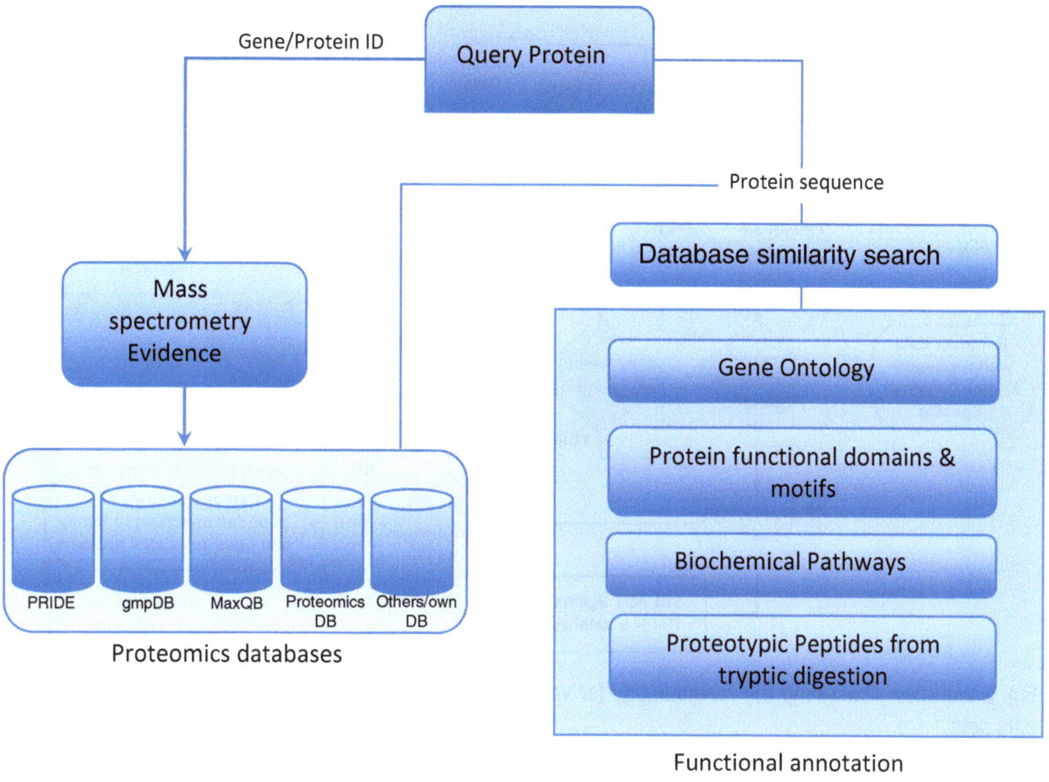

Fig. 1 The overall workflow for a systematic bioinformatics approach to identify high quality MS data and functionally annotate proteins and proteomes

annotations by mapping to protein domains, motifs and families complemented by gene ontology (GO) terms and biochemical pathways [10–12].

In this chapter, we provide a roadmap for a systematic bioinformatics approach to identify high quality MS data and functionally annotate proteins and proteomes, illustrated in Fig. 1. This analysis workflow addresses both MS data validation (Fig. 2) as well as functional annotation (Fig. 3). Firstly, we outline an intuitive approach to interpreting and validating MS data from various search engine software solutions using a simple workflow. We then describe a comprehensive functional annotation pipeline that is ideally suited to proteins and proteomes, with very little or no annotations available (e.g., novel organisms or "missing" proteins [10–12]). This pipeline is also suitable for reconfirming data obtained from proteomics experiments, before hypotheses are formulated. The methods we propose here would be a significant addition to any existing workflows in proteomics laboratories as it requires minimal computing power and results in biologically meaningful data interpretation.

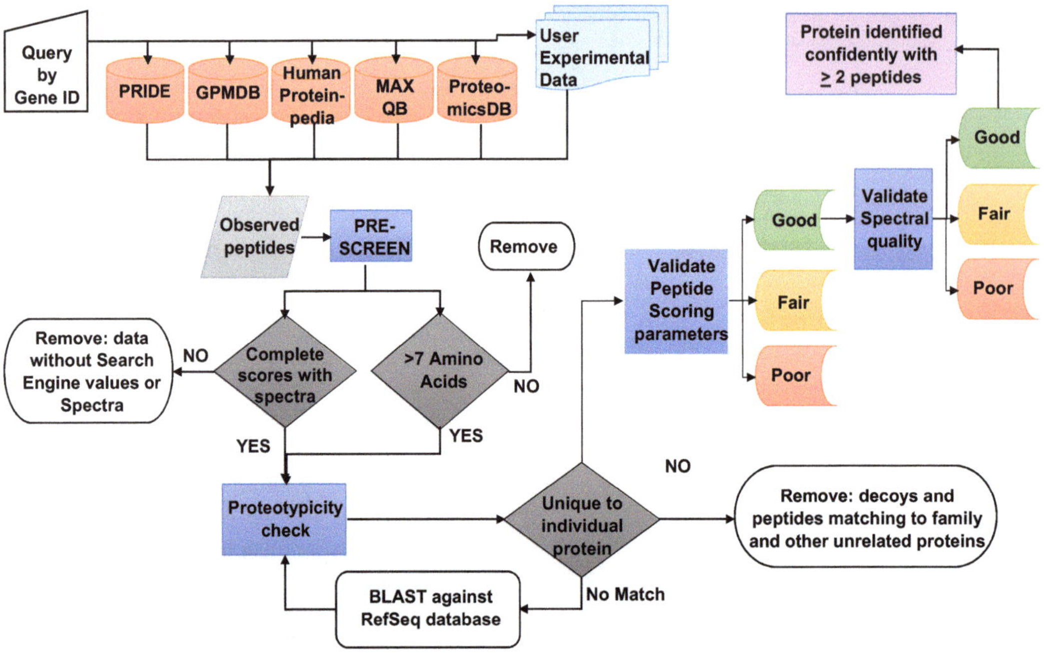

Fig. 2 Workflow illustrating the methodology for validation of MS data from repositories and user-generated data sets

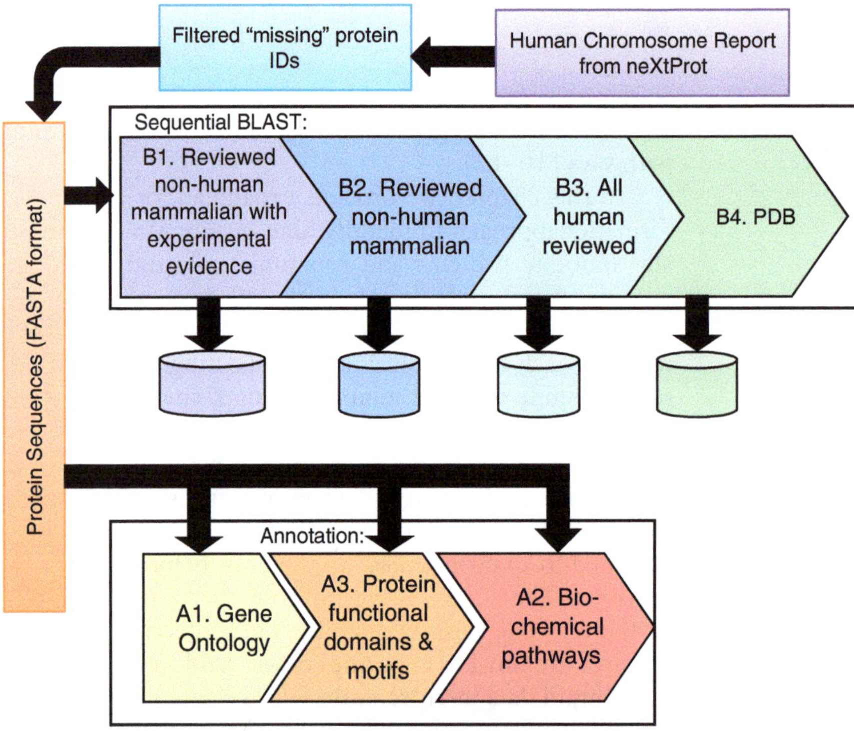

Fig. 3 Workflow illustrating the methodology for comprehensive protein functional annotation

2 Materials

2.1 Data Sources

2.1.1 MS Data Validation

MS data can be sourced from proteomics databases, listed in Table 1, for comparing with user-derived MS data. Users may derive data from any other relevant search engines or data repositories.

2.1.2 Functional Annotation

Retrieve all required information to your computer. For the human chromosome 7 example, the following are required:

1. The neXtProt [13] database report for human chromosome 7 (hChr7) [14];
2. Protein accession reports of protein evidence level 2–4 (transcript, homology and predicted) [15] from neXtProt [13] database;
3. FASTA sequences of hChr7 proteins with evidence level 2–4 from UniProt [16] (*see* **Note 1**);
4. Protein datasets in FASTA format from UniProt database [16]:
   (a) Non-human reviewed mammalian proteins with protein evidence,
   (b) Non-human reviewed mammalian proteins,
   (c) Human reviewed proteins;
5. Protein Data Bank (PDB) [17] proteins from the PDB [18].

2.2 Software

1. Online batch peptide match tool [19] for identifying proteotypic peptides;
2. BLAST [20] package of tools for database similarity search to identify proteins homologous to known proteins or structures [21];
3. InterProScan [22] for protein domain mapping [23];
4. KAAS [24] or KOBAS 2.0 [25] for pathway mapping.

Table 1
Publicly available proteomics databases

Name	URL
PRIDE	http://www.ebi.ac.uk/pride/archive/simpleSearch
gpmDB	http://gpmdb.thegpm.org
MaxQB	http://maxqb.biochem.mpg.de/mxdb/
Proteomics DB	https://www.proteomicsdb.org

Table 2
Relevant parameters collected from MS data analysis software, as collated by proteomics databases

Database	MS data analysis software	Information collected
1. PRIDE	Sequest Mascot Spectrum Mill Peptide Shaker	Sp, Xcorr Mascot Score Mascot Threshold Mascot Expectation Score Spectrum Mill Peptide Score Peptide Confidence Peptide Threshold PSM Confidence PSM Threshold
2. GPMDB	X! Tandem	*E*-value
3. Human Proteinpedia	–	Peptide Score
4. ProteomicsDB	Andromeda Mascot	Andromeda Score Mascot Score

3 Methods

3.1 MS Data Validation

Proteomic databases hold peptide information analyzed using different MS data analysis software by researchers worldwide. To demonstrate the efficacy of our method, we extracted peptide data from numerous repositories to demonstrate interpretation and analysis of data from different search engines (the process is outlined in Fig. 2). The results presented in this chapter are based on database entries as of May 2016. As the databases used for both data download as well as by the search tools in Section 2 are constantly updated, the exact peptide numbers may differ from those reported here.

1. Go the relevant webpages listed in Table 1, for PRIDE (either directly or through the link provided by UniProt for the protein) [26], GPMDB [27], MAXQB [28], and ProteomicsDB [29] databases and collect the MS/MS data for the human disintegrin and metalloproteinase domain-containing protein 8 (ADAM8) (Uniprot ID: P78325).
2. Collate all relevant information pertaining to the protein: analysis software (listed in Subheading 2.2) along with peptide sequence information (in this case, 148 peptides); availability of spectrum and number of observations (if provided by the database) for each peptide. Table 2 lists the information to be collected for each MS software program.
3. Preprocess the collected data by removing very short peptides (less than seven amino acids: six entries were removed) and

Table 3
Recommended values for MS data from the most commonly used search engine software

Search engine	Good	Moderate	Poor
Sequest Raw Sp Score	300	200	120
Raw XCorr Score	2	1.5	0.5
GPM X! Tandem *E* value	−10	−5.7	−3
Mascot Mascot Score Raw	7.5	4	2
Mascot Mascot Expectation Value	0.005	0.05	0.5
Spectrum Mill Spectrum Mill value	15	9	5
MAX-Quant Raw Andromeda Score	200	100	60
Peptide Shaker	PC > PT and PSMC > PSMT	PC > PT and PSMC < PSMT or PC < PT and PSMC > PSMT	PC < PT and PSMC < PSMT

PC peptide confidence, *PT* peptide threshold, *PSMC* PSM confidence, *PSMT* PSM threshold

peptides that have incomplete scores or no spectra (ten entries). This filtered dataset (132 peptides) is used for further analysis. Peptides of six amino acids or less are not considered significant for identification [9].

4. To avoid the protein inference problem [30], process the peptides through the online batch peptide match tool [19] using the UniProtKB protein database; select UniProt organism: Homo sapiens [9606] and ensuring that Leucine (L) and Isoleucine (I) are considered equivalent. Peptides with a single match to the exact protein ID (i.e., ADAM8) are proteotypic (seven peptides) and selected for further analysis. Peptides without a match in UniProtKB database can be manually searched against NCBI Reference Sequence (RefSeq) Database of humans (taxid: 9606) (*see* **Note 2**).
5. Review the MS software parameter scores for each peptide and rank them as "good" (50 peptides), "moderate" and "poor" (127 peptides combined) according to Table 3 (*see* **Note 3**).
6. Manually verify the correct peptide assignment for peptides with good scores by validating the spectra. Table 4 lists the

Table 4
Top scoring manually validated proteotypic peptides for ADAM8 from the different data repositories

	Database	Unique peptides of ADAM8 identified
1	GPMDB	GQDHCFYQGHVEGYPDSAASLSTCAG
2	GPMDB	AICIVDVCHALTTEDGTAYEPVPEGTR
3	GPMDB	GEQCDCGPPEDCR
4	GPMDB	GFFQVGSDLHLIEPLDEGGEGGR
5	GPMDB	CQDLHVYR
6	GPMDB	GDGAASRAGPL
7	GPMDB	SNPLFHQAASR
8	GPMDB	CIMAGSIGSSFPR
9	MaxQB	GPQEIVPTTHPGQPAR
10	MaxQB	PGAGAANPGPAEGAVGPK
11	MaxQB	VSAMCSHSSGAVNQDHSK
12	PRIDE	VRRALPSHLGLHPER
13	PRIDE	VKPAGELCR
14	PROTEOMICSDB	ADMCGVLQCK
15	PROTEOMICSDB	RPPPAPPVTVSSPPFPVPVYTR
16	PROTEOMICSDB	VKPAGELCRPK

peptides considered "good" from the example dataset (16 peptides) (*see* **Note 4**).

3.2 Sequential-BLAST Similarity Search

Use sequential database similarity search technique [12] to check if a target sequence is homologous to sequences that are already available in existing databases (*see* **Note 5**).

1. Download and install the latest BLAST [20] package of tools on your machine from [21].
2. Make BLAST-searchable databases for each of the reference databases mentioned in Subheading 2.1 [4, 5]. The command to create a BLAST database is:

```
makeblastdb -in<inputfile>-out<outputfile>-
dbtype prot
```

3. Perform a BLASTP search against the *Non-human reviewed mammalian proteins with protein evidence* database, using default parameters with a minimum E-value of 1*e*–05 to identify homologous proteins to target sequences. The command for carrying this out is:

```
blastp -num_threads<n>-query<input FASTA>-
db<path to blast database>-out<output file>-
evalue 1e-05 -outfmt 6
```

4. Sort your outputs according to target protein id, and sequence identity. Retain results with % identity value greater than or equal to 50. If you have multiple hits for the same protein sequence, retain the top hit only (*see* **Note 6**).
5. Compare your results with your input sequence list to identify sequences that yielded no matches in **step 3**. The sequences having no database match will be analyzed further.
6. Perform a BLASTP search against the *non-human reviewed mammalian proteins* dataset, use default parameters with a minimum E-value of 1*e*–0 5. Use the sequences from **step 5** as your input sequence.
7. Sort your outputs according to target protein id, and sequence identity. Retain results with % identity value greater than or equal to 50. If you have multiple hits for the same protein sequence, retain the top hit only (*see* **Note 6**).
8. Compare your results with your input sequence to identify sequences that yielded no matches in **step 6**. The sequences having no database match will be analyzed further.
9. Perform BLASTP search against the *non-human reviewed mammalian proteins* dataset, use default parameters with a minimum *E*-value of 1*e*–05. Use the sequences from **step 8** as your input sequence.
10. Sort your outputs according to target protein id, and sequence identity. Retain results with % identity value greater than or equal to 50. If you have multiple hits for the same protein sequence, retain the top hit only (*see* **Note 6**).
11. Compare your results with your input sequence to identify sequences that yielded no matches in **step 9**. The sequences having no database match will be analyzed further.
12. Perform BLASTP search against the *Human reviewed proteins* dataset, use default parameters with a minimum *E*-value of 1*e*–05. Use the sequences from **step 11** as your input sequence.
13. Sort your outputs according to target protein id, and sequence identity. Retain results with % identity value greater than or equal to 50. If you have multiple hits for the same protein sequence, retain the top hit only.
14. Compare your results with your input sequence to identify sequences that yielded no matches in **step 11**. Again, the sequences having no database match will be analyzed further.
15. Perform BLASTP search against the *PDB* dataset, use default parameters with a minimum *E*-value of 1*e*–05. Use the sequences from **step 14** as your input sequence.

16. Sort your outputs according to target protein id, and sequence identity. Retain results with % identity value greater than or equal to 50. If you have multiple hits for the same protein sequence, retain the top hit only (*see* **Note 6**).

3.3 Functional Annotation

3.3.1 Protein functional domains and motifs, and Gene Ontology (GO)

Use InterProScan [22] to assign putative functional annotation by mapping to protein domain, motif and families. This program can assign GO terms to the query proteins.

1. Open your browser and go to the InterproScan search page [31].
2. Copy and paste your protein sequence to the input box.
3. Select advanced search, then select the member databases for Families, domains, sites and repeats, and structural domains.
4. Click search, and wait for the process to complete (*see* **Note 7**).
5. Select export format (TSV) and download the data (*see* **Note 8**).

3.3.2 Pathway Analysis

Use KEGG Orthology-Based Annotation System (KOBAS -2.0) [25] for pathway mapping. This is a two-step process, first mapping the proteins to genes in KEGG GENES, based on BLAST searches to obtain pathway and disease annotations and then find enriched pathways and diseases against the human proteome as the background (*see* **Note 9**).

1. Open your browser and go to KOBAS-2.0 [25] webserver [32].
2. Click *Annotate* from the left hand menu.
3. Select input type as FASTA protein sequence.
4. Paste your sequence or upload a copy from your computer.
5. Choose the species or KO as *Homo Sapiens (Human)* (*see* **Note 10**).
6. Expand the Options for sequences or BLAST output and use *E-value* = *1e−5* and *BLAST subject coverage* = 0.50 (*see* **Note 6**).
7. Click run and wait for the process to complete.
8. Once completed, click on *use this file as the input to the Identify's input*, which will take you to the *Identify* processing page to identify enriched pathways, diseases and GO terms.
9. Run it with default parameters to select all databases.
10. Once completed, download the results to your local machine.

4 Notes

1. A simple script can be written to download FASTA sequences programmatically using freely available non-interactive command line tools such as Wget [33] and cURL [34].
2. Proteotypic peptides alone are used for protein identification. Peptides that uniquely match to an isoform or splice variant

could be used for further independent analysis and are filtered out in this high-stringency pipeline. Decoy peptides are also removed from data for downstream analysis. Some software solutions may integrate this step into the analysis but checking algorithms using this method are advised.

3. Current literature and manufacturer technical guidelines from the relevant vendors of the different data analysis software was collated and carefully studied to determine which parameter values could be considered "good," "moderate," and "poor" from the specific search engine software. From these parameter values, "good" scores were considered acceptable, while moderate scores could be used as supporting evidence for protein presence, while poor scores are of unacceptable confidence. These scores have to be associated usually with an FDR of <1 % at the protein level which can be set in the search engine parameters [35]. Database limitations may affect these results and hence using a comprehensive, well-annotated database such as UNIPROT is advisable.

4. Spectra and associated annotation, spectral match error values and fragmentation ions from most databases and software are readily available through the associated web graphical user interface. Spectra from the PRIDE database is not annotated and therefore individual PRIDE xml files need to be analyzed using the PRIDE Automatic Spectrum Annotation Pipeline [36]. The spectra have be visually assessed on a confidence-weighted scale of three primary criteria [37, 38] that are most important:

   (a) Spectral noise: values <0.3 Signal to noise ratio considered as little or no noise while values >0.3 signal to noise ratio would represent unacceptably large amounts of noise.

   (b) Error: where the assignment error was within 10 ppm/0.4 Da is considered acceptable, while values outside this threshold are unacceptable.

   (c) The run of singly charged ions: a good run of both b or y ions in the case of a CID experiment would be considered the best followed by a good run of either b or y ions, or run of other ions (x, z, Y++ or B++, etc.), whereas a haphazard run of ions would not be acceptable.

   Two secondary criteria that can be taken into consideration include the number of assigned peaks (for instance, having all major peaks assigned would be good whereas if only a small number of peaks are assigned, this spectrum may demonstrate a poor match) and relative intensity of the spectrum (major assigned peaks of moderate intensity (>20 %) or low intensity (<20 %). Recently, even a single unique proteotypic peptide at least 9aa long is considered sufficient to confidently identify a protein [9].

5. You can access the sample results for human "missing" proteins from ProtAnnotator [10] webserver [39].
6. We used *e*-value = 1*e*–5 and % identity = 50 for our studies. You can adjust these according to your experiment.
7. It is possible to download and install InterProScan [22] locally and run it with default parameters for batch processing however it requires computers with good memory and processing power and some Linux knowledge [40].
8. If you are using InterProscan webserver, disable the popup blocker on your browser to download data.
9. It is best to run this process as a registered user. Registered users can select the option to save their results in the working directory within the webserver. This will ensure your data not lost due to network outage or browser crash.
10. If you are analyzing nonhuman data, please select the species accordingly. Depending on your selections, databases will be displayed automatically.

References

1. Laukens K, Naulaerts S, Berghe WV (2015) Bioinformatics approaches for the functional interpretation of protein lists: from ontology term enrichment to network analysis. Proteomics 15(5-6):981–996. doi:10.1002/pmic.201400296
2. Kumar C, Mann M (2009) Bioinformatics analysis of mass spectrometry-based proteomics data sets. FEBS Lett 583(11):1703–1712. doi:10.1016/j.febslet.2009.03.035
3. Carnielli CM, Winck FV, Paes Leme AF (2015) Functional annotation and biological interpretation of proteomics data. Biochim Biophys Acta 1854(1):46–54. doi:10.1016/j.bbapap.2014.10.019
4. Draghici S, Khatri P, Martins RP, Ostermeier GC, Krawetz SA (2003) Global functional profiling of gene expression. Genomics 81(2):98–104. doi: 10.1016/S0888-7543(02)00021-6
5. Khatri P, Draghici S (2005) Ontological analysis of gene expression data: current tools, limitations, and open problems. Bioinformatics 21(18):3587–3595. doi:10.1093/bioinformatics/bti565
6. Goeman JJ, Buhlmann P (2007) Analyzing gene expression data in terms of gene sets: methodological issues. Bioinformatics 23(8):980–987. doi:10.1093/bioinformatics/btm051
7. Deutsch EW, Albar JP, Binz PA, Eisenacher M, Jones AR, Mayer G, Omenn GS, Orchard S, Vizcaino JA, Hermjakob H (2015) Development of data representation standards by the human proteome organization proteomics standards initiative. J Am Med Inform Assoc 22(3):495–506. doi:10.1093/jamia/ocv001
8. Haga SW, Wu HF (2014) Overview of software options for processing, analysis and interpretation of mass spectrometric proteomic data. J Mass Spectrom 49(10):959–969. doi:10.1002/jms.3414
9. Omenn GS, Lane L, Lundberg EK, Beavis RC, Nesvizhskii AI, Deutsch EW (2015) Metrics for the Human Proteome Project 2015: Progress on the Human Proteome and Guidelines for High-Confidence Protein Identification. J Proteome Res 14(9):3452–3460. doi:10.1021/acs.jproteome.5b00499
10. Islam MT, Garg G, Hancock WS, Risk BA, Baker MS, Ranganathan S (2014) Protannotator: a semiautomated pipeline for chromosome-wise functional annotation of the "missing" human proteome. J Proteome Res 13(1):76–83. doi:10.1021/pr400794x
11. Ranganathan S, Khan JM, Garg G, Baker MS (2013) Functional annotation of the human chromosome 7 "missing" proteins: a bioinformatics approach. J Proteome Res 12(6):2504–2510. doi:10.1021/pr301082p
12. Islam MT, Mohamedali A, Garg G, Khan JM, Gorse AD, Parsons J, Marshall P, Ranganathan

S, Baker MS (2013) Unlocking the puzzling biology of the black Perigord truffle Tuber melanosporum. J Proteome Res 12(12):5349–5356. doi:10.1021/pr400650c
13. Gaudet P, Argoud-Puy G, Cusin I, Duek P, Evalet O, Gateau A, Gleizes A, Pereira M, Zahn-Zabal M, Zwahlen C, Bairoch A, Lane L (2013) neXtProt: organizing protein knowledge in the context of human proteome projects. J Proteome Res 12(1):293–298. doi:10.1021/pr300830v
14. Full Chromosome Reports from neXtProt. ftp://ftp.nextprot.org/pub/current_release/chr_reports. Accessed 27 October 2016
15. Simplified chromosome reports from neXtProt. ftp://ftp.nextprot.org/pub/current_release/custom/hpp. Accessed 27 October 2016
16. UniProt Consortium (2012) Reorganizing the protein space at the Universal Protein Resource (UniProt). Nucleic Acids Res 40(Database issue):D71–75. doi:10.1093/nar/gkr981
17. Berman HM, Westbrook J, Feng Z, Gilliland G, Bhat TN, Weissig H, Shindyalov IN, Bourne PE (2000) The Protein Data Bank. Nucleic Acids Res 28(1):235–242. doi: 10.1093/nar/28.1.235
18. Protein Data Bank (PDB) http://www.rcsb.org/pdb/download/download.do. Accessed 27 October 2016
19. Chen C, Li Z, Huang H, Suzek BE, Wu CH (2013) A fast Peptide Match service for UniProt Knowledgebase. Bioinformatics 29(21):2808-2809. doi: 10.1093/bioinformatics/btt484
20. Altschul SF, Gish W, Miller W, Myers EW, Lipman DJ (1990) Basic local alignment search tool. J Mol Biol 215(3):403–410. doi:10.1016/S0022-2836(05)80360-2
21. NCBI BLAST ftp://ftp.ncbi.nlm.nih.gov/blast/executables/blast+/LATEST/. Accessed 27 October 2016
22. Quevillon E, Silventoinen V, Pillai S, Harte N, Mulder N, Apweiler R, Lopez R (2005) InterProScan: protein domains identifier. Nucleic Acids Res 33 (Web Server issue):W116-120. doi:10.1093/nar/gki442
23. InterProScan. http://www.ebi.ac.uk/Tools/pfa/iprscan5/ http://www.ebi.ac.uk/interpro/search/sequence-search. Accessed 27 October 2016
24. Moriya Y, Itoh M, Okuda S, Yoshizawa AC, Kanehisa M (2007) KAAS: an automatic genome annotation and pathway reconstruction server. Nucleic Acids Res 35 (Web Server issue):W182-185. doi:10.1093/nar/gkm321
25. Xie C, Mao X, Huang J, Ding Y, Wu J, Dong S, Kong L, Gao G, Li CY, Wei L (2011) KOBAS 2.0: a web server for annotation and identification of enriched pathways and diseases. Nucleic Acids Res 39 (Web Server issue):W316-322. doi:10.1093/nar/gkr483
26. Martens L, Hermjakob H, Jones P, Adamski M, Taylor C, States D, Gevaert K, Vandekerckhove J, Apweiler R (2005) PRIDE: the proteomics identifications database. Proteomics 5(13):3537–3545. doi:10.1002/pmic.200401303
27. Craig R, Cortens JP, Beavis RC (2004) Open source system for analyzing, validating, and storing protein identification data. J Proteome Res 3(6):1234–1242. doi:10.1021/pr049882h
28. Schaab C, Geiger T, Stoehr G, Cox J, Mann M (2012) Analysis of high accuracy, quantitative proteomics data in the MaxQB database. Molecular & cellular proteomics : MCP 11 (3):M111 014068. doi:10.1074/mcp.M111.014068
29. Wilhelm M, Schlegl J, Hahne H, Gholami AM, Lieberenz M, Savitski MM, Ziegler E, Butzmann L, Gessulat S, Marx H (2014) Mass-spectrometry-based draft of the human proteome. Nature 509(7502):582–587. doi: 10.1038/nature13319
30. Nesvizhskii AI, Aebersold R (2005) Interpretation of shotgun proteomic data: the protein inference problem. Molecular & cellular proteomics : MCP 4(10):1419–1440. doi:10.1074/mcp.R500012-MCP200
31. InterProScan Search. http://www.ebi.ac.uk/interpro/search/sequence-search. Accessed 27 October 2016
32. KOBAS 2.0. http://kobas.cbi.pku.edu.cn. Accessed 27 October 2016
33. Scrivano G GNU Wget. http://www.gnu.org/software/wget/. Accessed 27 October 2016
34. Stenberg D curl. http://curl.haxx.se/. Accessed 27 October 2016
35. Deutsch EW, Sun Z, Campbell D, Kusebauch U, Chu CS, Mendoza L, Shteynberg D, Omenn GS, Moritz RL (2015) State of the Human Proteome in 2014/2015 As Viewed through PeptideAtlas: Enhancing Accuracy and Coverage through the AtlasProphet. J Proteome Res 14(9):3461–3473. doi:10.1021/acs.jproteome.5b00500
36. Hulstaert N, Reisinger F, Rameseder J, Barsnes H, Vizcaino JA, Martens L (2013) Pride-asap: automatic fragment ion annotation of identified PRIDE spectra. Journal of proteomics 95:89–92. doi:10.1016/j.jprot.2013.04.011
37. Sadygov RG, Cociorva D, Yates JR 3rd (2004) Large-scale database searching using tandem mass spectra: looking up the answer in the back of the book. Nature methods 1(3):195–202. doi:10.1038/nmeth725

38. Craig R, Beavis RC (2004) TANDEM: matching proteins with tandem mass spectra. Bioinformatics 20(9):1466–1467. doi:10.1093/bioinformatics/bth092
39. Protannotator. http://www.biolinfo.org/protannotator/human_Chr7.php. Accessed 27 October 2016
40. InterProScan Download and Requirements. https://github.com/ebi-pf-team/interproscan/wiki/HowToDownload AND https://github.com/ebi-pf-team/interproscan/wiki/InstallationRequirements. Accessed 27 October2016

Chapter 14

Network Tools for the Analysis of Proteomic Data

David Chisanga, Shivakumar Keerthikumar, Suresh Mathivanan, and Naveen Chilamkurti

Abstract

Recent advancements in high-throughput technologies such as mass spectrometry have led to an increase in the rate at which data is generated and accumulated. As a result, standard statistical methods no longer suffice as a way of analyzing such gigantic amounts of data. Network analysis, the evaluation of how nodes relate to one another, has over the years become an integral tool for analyzing high throughput proteomic data as they provide a structure that helps reduce the complexity of the underlying data.

Computational tools, including pathway databases and network building tools, have therefore been developed to store, analyze, interpret, and learn from proteomics data. These tools enable the visualization of proteins as networks of signaling, regulatory, and biochemical interactions. In this chapter, we provide an overview of networks and network theory fundamentals for the analysis of proteomics data. We further provide an overview of interaction databases and network tools which are frequently used for analyzing proteomics data.

Key words Proteomics, Network theory, Protein–protein interactions, Network tools, Network analysis, Bioinformatics

1 Introduction

In recent years, the development of high-throughput technologies such as next-generation sequencing techniques in the field of genomics and tandem mass spectrometry in the field of proteomics and metabolomics has led to the birth of the "omics" study [1]. These techniques and tools involved in the study of functional genomics and other omics data have constantly helped in our understanding of cellular biology and have drastically reduced the cost of conducting "omics" related studies. The speed with which data are generated and disseminated today means that researchers can gain insight for the fraction of the cost compared to that in past years. For instance, by using tandem mass spectrometry, two groups [2, 3] have developed the first draft of the human proteome. Also, using bioinformatics, another group integrated

Shivakumar Keerthikumar and Suresh Mathivanan (eds.), *Proteome Bioinformatics*, Methods in Molecular Biology, vol. 1549, DOI 10.1007/978-1-4939-6740-7_14, © Springer Science+Business Media LLC 2017

publicly available proteomics datasets to map 96% of the human proteome [1].

However, with terabytes of proteomic data pouring into research centers every day, standard statistical methods for analyzing data are becoming ineffective. Researchers are faced with the formidable task of how to take advantage of this heterogeneous data to gain insight in areas such as disease and drug development as well as answering questions such as the following: How can they characterize and manipulate complex interactome of basic elements such as genes and proteins? How can they visualize these interactomes and infer meaningful information from them?

Network theory has long played a fundamental role in disciplines ranging from computer science, sociology, engineering, and physics, to molecular and population biology [4]. In biology and medicine, network analysis methods are applied in areas such as drug target identification, prediction of a gene or protein function, protein complex or module detection, prediction of novel interactions and functional associations, identification of disease subnetworks, disease biomarker identification, and mapping of disease pathways [5]. Networks have long been used in a variety of fields to reduce the complexity of data [6, 7]. Computational tools, including pathway databases and network building tools, have been developed to store, analyze, and interpret biological networks [8].

This chapter provides an overview of the application of network theory in analyzing and visualization of proteomic data by discussing various tools used for storage, analysis, and interpretation of proteomic data through the use of biological networks with an emphasis on protein–protein interaction networks. To get started, we provide a brief background to both proteomics and network theory.

1.1 Background to Proteomics

Coined by Marc Wilkins and colleagues [9] in the mid-1990s to mimic the terms "genomics" and "genome," respectively, proteomics is in essence a systems science whose aim is to identify and record the functions as well as structures of proteins in organisms. Proteomics is a systems science which involves not only the measurement of proteins but also the measurement of their expressions in a cell and the interplay of proteins, protein complexes, signaling pathways, and network modules.

Proteins are termed as the workhorses of cellular systems, as they perform an array of cellular functions ranging from catalyzing reactions, cellular transportation, transcription of DNA information to RNA, and acting as molecular motors to signaling [10]. They perform these functions not on their own, but within large complexes where they interact with other molecules like proteins, DNA, RNA as well as with other small molecules. Because of their importance, a malfunction in key proteins can lead to serious pathological outcomes like cancer, metabolic imbalances, and neurodegenerative diseases. With significant ongoing research into protein functionality and their interactions with other molecules in understanding disease, research has turned to network theory concepts to model and study these interactions.

1.2 Background to Network Theory Concepts

A network or a graph (in mathematics) is a collection of objects connected by lines. The objects are called nodes or vertices while the connections between the objects are called edges or links and are drawn as lines between points as shown in Fig. 1

Formally, a network is a graph G defined as an ordered pair $G=(V, E)$ where V is a set of nodes and E is a set of edges [4]. Nodes are said to be adjacent if they are joined by an edge while node 'A' is said to be a neighbor to node 'B' if adjacent to node 'B' and vice versa. Edges between nodes can be undirected (Fig. 1) or directed (Fig. 2), as such a graph $G=(V, E)$ is called undirected if

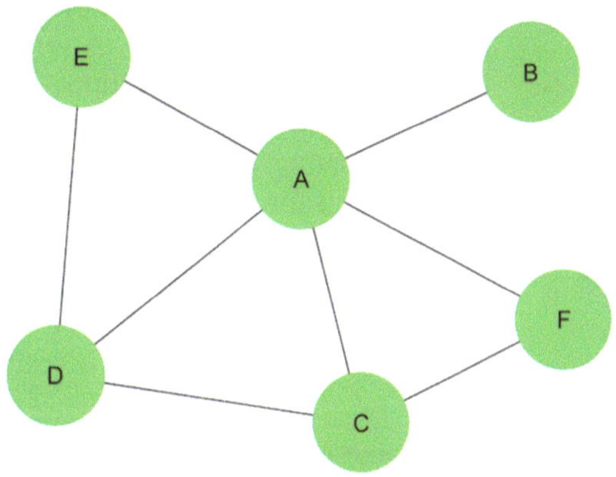

Fig. 1 Shows an example of an undirected network graph in which each node is connected by an edge that does not show the origin and destination by way of an ***arrow***

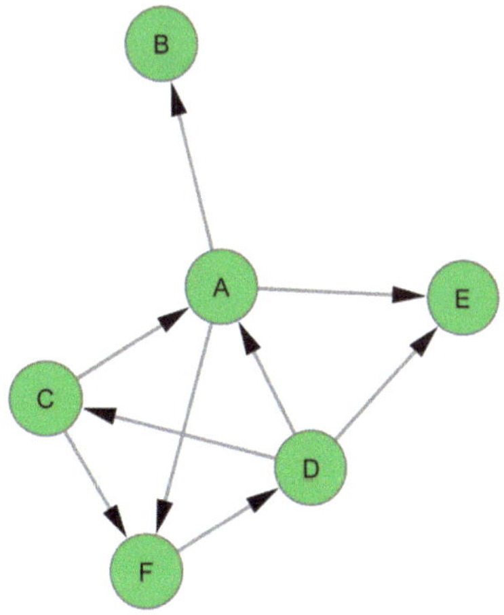

Fig. 2 Shows an example of a directed graph in which each node is connected by an edge with an ***arrow*** indicative of the direction of the relationship

an edge vv' (where v and v' are nodes in set *V*) in set E of edges implies that it is the same as edge v'v also in *E*; otherwise *G* is called directed. A directed acyclic graph, on the other hand, is a directed graph that contains no cycles. Finally, a graph is said to be connected if there is a path from any node to any other node.

Using the above network/graph concepts, researchers have used networks to reduce the complexity of systems thereby making it easier to draw conclusions from them. Networks are applied in various fields such as computer networks, social networks, and interactome networks in molecular biology research.

Interactome networks provide a global picture that is useful in understanding how interactions between molecules influence cellular behavior [11]. It has been established that biological behavior arises from the complex interactions between the cell's numerous molecules such as proteins, DNA, RNA, and other small molecules. Common examples of interactomes in molecular biology are; protein–protein interactions, virus–host networks, transcriptional regulatory networks, metabolic networks, and disease networks. Protein–protein interactions (PPIs) form the backbone of signaling pathways, metabolic pathways, and cellular processes required for normal functioning of cells [12].

The steps to perform proteomic analysis can be summed up by use of a flowchart as shown in Fig. 3, it involves identifying a set of target proteins of biological interest needs to be studied and then followed by retrieval or identification of interacting partners from

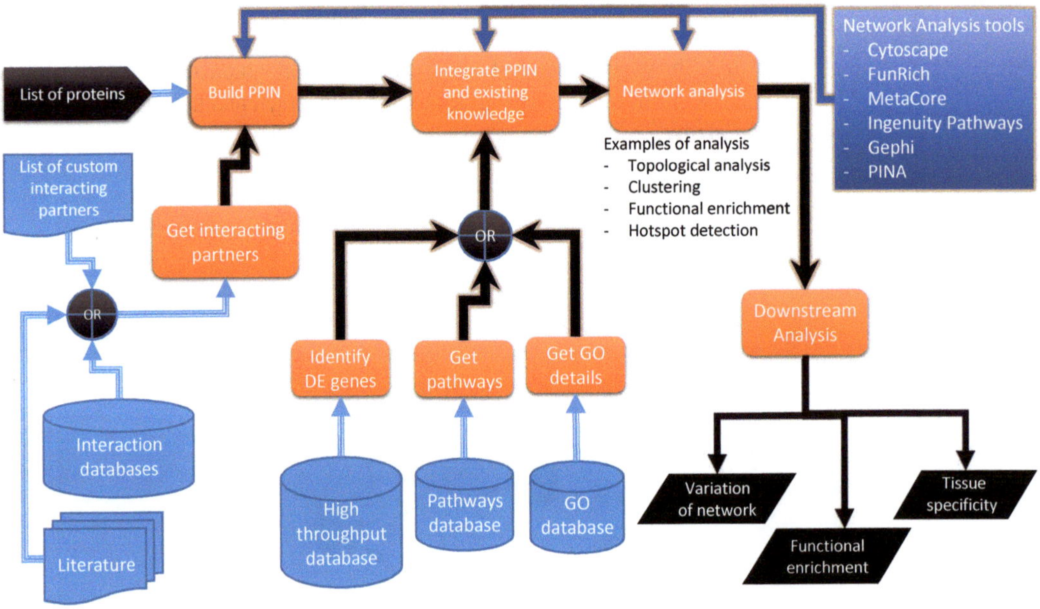

Fig. 3 Shows a summary representation of the steps involved in analyzing proteomic data using network theory concepts. It also shows the data types required and from where they can be sourced. It also gives an example of expected outputs from the network analysis

various interaction resources discussed below. An interaction network is then generated and integrated with any existing knowledge such as gene ontology (GO) enrichment, biological pathways or differential gene or protein expression. A topological analysis of the network is then performed using metrics such as degree, degree centrality or betweenness centrality which is further followed on by downstream analysis to identify network variations, functional enrichment of identified modules, or tissue specificity.

2 Protein–Protein Interaction Databases

The mappings of proteins and their interacting partners have been curated by various groups and deposited into online databases. These databases are typically Web-based resources that serve as archives of information pertaining to the mapping of protein interactions, functional enrichment (GO enrichment) and pathway details. These databases act as sources of protein mapping information in network analysis. The most widely used PPI databases include Human Protein Reference Database (HPRD) [13], Molecular Interaction Database (MINT) [14], Biological General Repository for Interaction Database (BioGRID) [15], Search Tool for Recurring Instances of Neighboring Genes/Proteins (STRING) [16], Database of Interacting Proteins (DIP) [17], Biomolecular Interaction Network Database (BIND) [18], and the IntAct molecular interaction database (IntAct). Depending on the database, the annotations may be based on experimental observations while other databases such as STRING can have a high proportion of predicted and literature mined interactions. Below, we briefly discuss the most commonly used databases while Table 1 provides a summary of these database resources with protein–protein interaction mappings.

2.1 BioGRID

The Biological General Repository for Interaction Datasets (BioGRID) is an open, accessible Web-based repository of genetic and protein interaction mappings which have been curated from the primary biomedical literature of humans and other major model organism species [15]. As of May 2016, the database housed over 1,000,000 protein and genetic interactions curated from over 56,000 high-throughput datasets and individually focused publications for major model organisms.

BioGRID features an easy to use Web interface with a search tool which users can use to search against the database, the search results then show the interacting partners, interactor details, and a graphical network visualization of the interacting partners. Users can then manipulate the network by either changing the network layout or filtering through the network by node degrees. In addition, users can also download custom defined or entire interaction

Table 1
Summary of database resources that house protein–protein interactions and their respective features

Resource	Description	URL link	Reference	No. proteins	No. interactions	No. organisms
BIND	Biomolecular Interaction Network Database	http://bond.unleashedinformatics.com/	[18]	23,643	43,050	80
BioGRID	Biological General Repository for Interaction Datasets	http://thebiogrid.org/	[15]	56,105	553,827	175
HPRD	Human Protein Reference Database	http://www.hprd.org	[13]	30,047	41,327	1
IntAct	IntAct Molecular Interaction Database	http://www.ebi.ac.uk/intact/	[20]	89,716	356,806	131
MINT	Molecular INTeraction database	http://mint.bio.uniroma2.it/mint	[14]	35,553	241,458	144
STRING	Search Tool for the Retrieval of Interacting Genes/Proteins	http://string-db.org/	[16]	9,643,763		2031

datasets for offline network analysis and downstream analysis. BioGRID also features online tools and resources that allow for the use of BioGRID data. A number of visualization tools such as Osprey, Cytoscape, and GeneMania, data management tools like ProHits, plugins like BioGRID Tab File Loader Plugin for Cytoscape and BiogridPlugin2 for Cytoscape as well as Web services BioGRID REST Service and PSICQUIC provide users with access to or can be used to analyze BioGRID data.

2.2 Human Protein Reference Database

Human Protein Reference Database is a Web-based resource that houses experimentally derived human proteome information [13]. It is one of the most comprehensive collections of human proteome information resource available online. It houses information pertaining to; protein–protein interactions, posttranslational modifications and tissue expression. As of May, 2016, the database housed over 30,000 protein entries, over 41,000 protein–protein interactions, 93,000 posttranslational modifications (PTMs), 112,000 protein expressions, 22,000 subcellular localization details, 400 domains and with over 453,000 PubMed links to publications.

The landing page of HPRD provides a range of features ranging from a querying functionality, BLAST feature to a browse feature. Users can query the database using the query page through a number of search options, the results are then displayed using graphical visual displays and are categorized into protein information, PTMs, protein length, and protein–protein interactions. Users can similarly get protein information through the browse page where the information is grouped into molecular classes, domains, motifs, PTMs and based on localization. HPRD further includes a Basic Alignment Search Tool (BLAST) which allows users to search against the database based on the provided protein or nucleotide sequence. Other features included are a phosphor motif finder tool which searches across user submitted protein sequence for the presence of over 300 phosphorylation-based motifs listed in HPRD. HPRD also provides tab delimited files for binary protein–protein interactions which users can download for offline processing and further download stream analysis.

2.3 Molecular INTeraction Database (MINT)

The Molecular INTeraction database [19] is a Web-based resource that stores physical interactions between proteins of model organisms that have been curated from the scientific literature. As of May 2016, MINT had over 241,000 protein–protein interactions, 35,000 proteins, and over 5000 PubMed links to publications.

MINT data can be downloaded in several formats such as PSI-ML, tab-delimited and MINT flat file formats. Otherwise, users can use the search feature that allows users to search the MINT database. Users can search the database using several options such as by gene name, protein accession number, or any 6-character keyword. A user defined list of proteins can furthermore be uploaded and used to generate a network visualization based on the information in the database.

2.4 Biomolecular Interaction Network Database

The Biomolecular Interaction Network Database [18] is a Web-based resource for PPI data and was one of the earliest resources for biomolecular interactions (proteins, genes, etc.), molecular complexes and pathways. BIND initiated by the University of Toronto as part of the Biomolecular Object Network Databank (BOND) has since been acquired by Thomson Reuters. BIND provides tools for data specification plus a database which is accompanied by data mining and visualization tools.

2.5 IntAct Molecular Interaction Database

IntAct [20] is an open-source Web-based molecular interaction database that catalogs data curated from the scientific literature or from direct data depositions. As of May 2016, IntAct had over 591,000 molecular interactions, and 91,000 interactors sourced from over 14,000 publications.

Using IntAct users can explore the fine details of the mechanism by which a specific protein binds to protein partners or use the entire interactome of an organism to perform a network analysis of large-scale 'omics' experiment. The front-end of IntAct features a search tool that can be used to search against the IntAct database. Users can then view the interacting partners, interaction details and a graphical presentation of the network.

2.6 Search Tool for Recurring Instances of Neighboring Genes/ Proteins (STRING)

STRING is a freely available Web-based biological database that houses information on experimentally derived and predicted protein–protein interactions for a number of organisms. This information has been curated from various sources, including experimental data, computational prediction methods, and published literature. STRING holds over 184 million interactions, 9,000,000 proteins from over 2000 organisms.

STRING provides an easy-to-use Web interface that allows users to quickly search for a protein of interest and visualize and download interaction data. It further has a Cytoscape plugin which allows users to directly access the STRING database from Cytoscape. The interaction data returned from STRING is weighted and allows for the calculation of confidence scores for each interaction. In addition, STRING has capabilities that allow it to connect to other databases and consequently perform literature mining. It also includes a capability that allows for the drawing of simple protein networks based on the provided list of genes and the available interactions in the database.

3 PPI Data Exchange Formats

Interaction networks are represented in a number of different file formats, the most widely used formats are; tab delimited text (.tab or .txt format), excel workbooks (.xls format), simple interaction file (SIF or .sif format), nested network format (NNF or .nff format), graph markup language (GML or .gml format), XGMML (extensible graph markup and modeling language), SBML, BioPAX, PSI-MI level 1 and 2.5 formats. All the interaction repositories provide at least one of these formats as a way to download interaction data.

3.1 Delimited Text and Excel Workbooks

The delimited text and excel workbook file formats are the most basic and widely used for working with interactive data and are supported by most if not all network analysis tools. Tables in these files can contain network and edge (interaction) attributes or values such as the confidence of an interaction. With these types of files, users can specify the columns for source and target nodes as well as interaction types, and edge attributes when importing network data into an analysis tool.

3.2 Simple Interaction Format (SIF)

This format allows for the construction of a network from a list of interactions by easily merging different interaction sets into a larger network.

Each line of an SIF file annotates a source node, a relationship (or edge type), and one or more target nodes as shown in the following example:

```
nodeA <relationship type> nodeB
nodeC <relationship type> nodeB
nodeD <relationship type> nodeA
```

3.3 Nested Network Format

This format is simple and similar to the SIF format except it allows the option of nesting a network into a single a node. An interaction is specified by either of two possible formats [21, 22]:

- A node "node" contained in a "network":
  - Network node.
- Two nodes linked together contained in a network:
  - Network node1 interaction with node2.

3.4 Graph Markup Language (GML)

GML unlike the SIF format comes with a language that supports rich graph formatting and is widely supported by most visualization software tools. A GML formatted file can contain information pertaining to graphs, nodes, and edges, and hence capable of emulating almost every other format. A network can be built using the SIF format and by applying network layouts can then be stored as a GML file as this preserves the layout of a network. Further details on the GML specification can be found on the GML documentation website: http://www.fim.uni-passau.de/index.php?id=17297&L=1.

Other formats such as XGMML is the XML extension of the GML format and is the preferred format to GML, Systems Biology Markup Language (SMBL) format is an XML format used to describe biochemical networks, the specification for SMBL can be found on the website: http://sbml.org/Documents/Specifications, PSI-ML format specification is an XML-based format that is used for data exchange of protein–protein interactions. GraphML is another XML-based format for generating graphs. Apart from the XML-based formats, JSON-based file formats are increasingly being used for data exchange of protein–protein interactions (Subheading 2.3).

4 Network Analysis and Visualization Tools

This section discusses some of the commonly used tools in the proteomics network analysis, but before delving into what tools to use, we begin this discussion by looking at the ways by which networks can be quantified in order to provide more informative results.

4.1 Quantifying Networks

The most commonly applied metric are; degree, degree distribution, scale-free networks, the degree exponent, shortest path, mean path length, and clustering coefficient [23]. By using these network metrics, we can quantify and characterize important network features which are not commonly visible.

Protein–protein interactions are the most commonly used form of networks in proteomic data analysis. In these networks, proteins are represented as nodes while interactions between the nodes are depicted by edges or links. This mapping of proteins is based on experimental information which has been obtained from methods such as mass spectrometer [24], protein chip technologies [25, 26], yeast two-hybrid screens [27], and predictions from computational methods [28]. These mappings have been collected and deposited into online databases as discussed below.

Network tools are mainly used to analyze proteomic data through functional annotation, knowledge integration, modularity analysis, topological analysis, and basic network property analysis [29].

The basic properties of a network such as node degree, degree distribution, betweenness centrality, and eigenvector centrality can be used to deduce the significance of a protein [30]. Another important metric is the identification of modules which represent a vital level of organization in biology [31]. A module in proteomics can be defined as a set of interacting proteins that can be associated with a common biological process. By using networks, clusters of interacting proteins can be identified as modules and associated with a functionality. Modules provide a comprehensive and global description of interaction patterns to comprehend the complexity of biological systems [32]. Module detection enables functional annotation of constituent proteins and the discovery of targets for therapy in diseases such as cancer. In addition to detection of modules, the integration of existing knowledge into networks plays a vital role in the analysis of proteomic data. Such knowledge may include integrating Gene Ontology (GO) annotations, differential gene expression, and pathway details. By highlighting such information, candidate disease proteins may be identified and module functions can be annotated.

4.2 Steps to Performing Network Analysis

To perform network analysis on proteomic data, there are a number of steps that are involved; these steps are summarized in Fig. 3. The steps involved include but are not limited to:

1. The first step involves identifying a list of proteins or genes that need to be analyzed using a network tool. The researcher can select which protein or gene appears on the lists, as per individual needs.
2. Interacting partners of these proteins are then obtained from any of the databases discussed above.
3. A protein–protein interaction network is then built by using a visualizing tool from the tools listed in Table 2.
4. To get more meaningful information from the network, the protein–protein interaction network is then integrated with already existing knowledge such as pathways, differential expressions for genes or proteins obtained from either high-throughput custom data or online databases such as The Cancer Genome Atlas (TCGA). Other existing knowledge that can be integrated includes Gene Ontology enrichment, which can help to identify the functional annotations of the modules or individual proteins in the network.
5. During topological analysis, network theory concepts such as degree, degree centrality distribution, Eigenvectors, and degree distribution are applied to identify proteins or nodes playing significant roles in the network, variations between a normal and an altered network and modules that can be mapped to a functionality.
6. Topology analysis is further followed by downstream analysis whose objective is mostly dependent on the researcher.
7. Some of the results that may be obtained from a network analysis of proteomic data include a visual representation of the network, module identification, network variations as well as functional enrichment of proteins and modules.

Table 2
Summary of Network tools for analyzing proteomic data

Tool	Reference	URL link	Features
Cytoscape	[22]	http://cytoscape.org/	Open source, Data integration, Network visualization, Network Analysis, Functional enrichment, extensible by plugins, Stand-alone, Platform independent
FunRich (Functional Enrichment Analysis)	[8]	http://funrich.org/	Open source, Functional enrichment, Dataset comparison, Network visualization and analysis, Stand-alone, Runs only on Windows, Results can be exported in various formats
MetaCore	By Thomson Reuters	https://portal.genego.com/	Proprietary, Network visualization, Network analysis, Function enrichment analysis, Data mining toolkit, Network alignment
Ingenuity Pathways Analysis	IPA®, QIAGEN Redwood City	www.qiagen.com/ingenuity	Proprietary, Network visualization and modeling, Causal network analysis, Network analysis, Functional enrichment analysis, Pathway enrichment analysis, Literature mining, Allows for collaboration
Gephi	Gephi	https://gephi.org	Network visualization, Network analysis, Network clustering, Module identification, Dynamic network analysis, Real-time visualization
PINA: Protein Interaction Analysis	[37]	http://cbg.garvan.unsw.edu.au/pina/	Network construction, Module detection, Functional enrichment, Network metric analysis, Network visualization, Community driven annotation
Osprey	[39]	http://biodata.mshri.on.ca/osprey/servlet/Index	Network visualization, Integrates BioGRID, Ability to compare functions between datasets, Build interaction network from custom datasets, Search for specific genes within a network, filtering feature

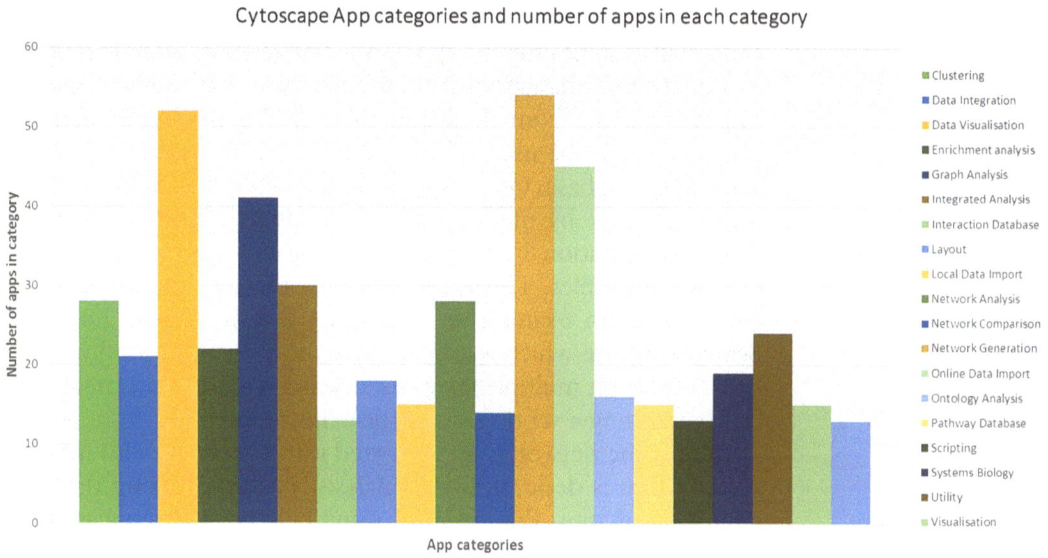

Fig. 4 Shows the distribution of apps or plugins across a number of categories in Cytoscape

4.3 Cytoscape

Cytoscape developed by Trey Ideker (a leading pioneer of systems biology) is a platform independent and open source software tool for the integration, visualization, and statistical modeling of molecular networks together with other systems-level data [21, 33]. The core of Cytoscape provides users with the fundamental features to perform functions such as data integration, analysis, and network visualization. The core also has limited information stored but interconnects with other databases to obtain relevant information. Cytoscape functionality is extensible through the integration of plugins (http://apps.cytoscape.org/) which are now called apps from version 3.0 of Cytoscape.

The apps can be categorized into one or more of the following functional categories such as clustering, data integration, data visualization, enrichment analysis, graph analysis, and integrated analysis. Other functional categories include interaction database, layout, local data import, network analysis, network comparison, network generation, online data import, ontology analysis, pathway database, scripting, systems biology, utility, and visualization. Figure 4 shows the distribution of these apps across the different functional categories.

The first step to a typical Cytoscape workflow is the importation of interactions. These interactions are imported from either a user's own experiment data or from public databases. Data from experiments is loaded directly into Cytoscape using a standard file format such as generic tabular formats including CSV, Excel, and TSV or network-specific formats such as SIF, XGMML, GML, PSI-MI, BioPAX (Biological Pathway Exchange), OpenBEL (Open Biological Expression Language), and SBML.

Importation of data from databases, on the other hand, requires the installation of plugins (apps). A list of genes of interest is passed as a query for interactions from the database. Examples of apps for importing data from databases include the BioGRID database plugin that can be used to import an entire interactome from the BioGRID database. Other ways in which networks can be imported into a network by mining interactions directly from literature or using computational inference from non-interaction data such as expression profiles. This is also achieved through the use of third-party apps. An example of such apps that is Agilent Literature Search software which is a meta-search tool that can automatically search through multiple texts based search engines to extract associations among a set of genes or proteins of interest.

Once the networks are imported into Cytoscape and network visualization is done, network analysis is achieved using the huge collection of apps. For example, using network topology apps like Knowledge-fused Differential Dependency Network (KDDN), users are able to calculate such statistics as network distribution of node degrees. Network clustering apps such as MCODE enable users to extract network regions which are densely connected, thereby forming modules which can then be related to complexes or pathways. Network enrichment apps are used to infer the functions of the identified modules by detecting functional terms that are statistically overrepresented among the set of genes making up the module. Examples of apps that can perform functional enrichment include BiNGO which is a tool that can determine which Gene Ontology categories are statistically overrepresented in a set of genes or a module, the ReactomeFIPlugin is another app that can be used to associate a set of genes in a module to pathways that are related to diseases such as cancer. Furthermore, functional modules can also be identified by integrating networks with expression data to infer network regions that are consistently up- or downregulated. Another example of network analysis that can be done using apps in Cytoscape is network comparison, this involves comparing networks across species or in different conditions to identify regions of the network with conserved interactions. GASOLINE (Greedy and Stochastic algorithm for Optimal Local Alignment of Interaction NEtworks) is an example of an app that can be used to compare multiple networks.

Cytoscape also supports the use of scripting languages such as Python and R. It enables users to develop their own scripts and integrate or call Cytoscape functionality in the order they want it to be done.

4.4 FunRich

Functional Enrichment Analysis (FunRich) tool [8] is an open source stand-alone desktop software tool for functional enrichment and protein–protein interaction network analysis of biological molecules. Features of FunRich include functional enrichment and network analysis of genes and proteins. In addition, FunRich allows the representation of results in editable graphical form as Venn, Bar, Column, Pie and Doughnut charts. FunRich users can perform a biological process, cellular component, molecular function, protein domain, site of expression, biological pathway, transcription, and clinical synopsis phenotypic term enrichment. Users can analyze their datasets against two built-in background databases; FunRich and UniProt or against a customized background database. FunRich does not require users to install any additional applications or plugins to conduct any of the above analysis. FunRich is currently only available for Microsoft's Windows Operating system with plans underway to support other major operating system platforms.

The first step to performing an enrichment analysis in FunRich is the specification of an annotation database. By default, FunRich comes with a human annotation database. Each database consists of biological function annotations and an interaction database. FunRich also comes with the latest UniProt annotation database, otherwise, users can also include a custom database. Once an annotation database has been specified, a list of genes or proteins is then imported. The user can perform a range of analyses on the datasets including comparison across the datasets using a Venn diagram that shows which proteins or genes are common across the datasets. Users can also perform gene set enrichment analysis to determine what biological functions are statically enriched in the gene or protein lists. In addition to these, FunRich also allows users to generate and build an interaction network from where users can then manipulate the network through enriched pathways and functions.

4.5 MetaCore

MetaCore from Thomson Reuters is an integrated proprietary software suite capable of analyzing multiple types of biological data, for example, Next Generation Sequencing [34], variant, Copy Number Variation (CNV), microarray, metabolic, proteomics, microRNA etc. Functional analysis in MetaCore is performed against a high quality, a manually curated database containing molecular interactions vis-à-vis protein–protein interactions, protein–DNA interactions, and protein–RNA interactions. The database is also made up of molecular classes such as transcription factors, signaling and metabolic pathways, and disease ontologies. MetaCore was developed for the purpose of representing biological functionality along with the integration of functional, molecular, or clinical information. Using the data mining toolkit available in MetaCore, users can perform functions like data visualization, analysis, and exchange of data, network alignment using multiple network alignment algorithms, and enrichment analysis. While MetaCore provides a set of rich features, it is a paid for a suite of software for integrated analysis.

4.6 Ingenuity Pathways Analysis

IPA (IPA®, QIAGEN Redwood City, www.qiagen.com/ingenuity) is a proprietary software application with features that allow scientists to model, analyze, and understand the complexity of biological and chemical systems [35]. IPA offers a host of network analysis functions some of these include causal network analysis which allows researchers to identify upstream molecules that control the expression of genes in their datasets and network analysis which allows the building and exploration of transcription of molecular networks such as microRNA, transcriptional networks, and protein–protein interaction networks. Network analysis in IPA can identify regulatory events that lead from signaling events to transcriptional effects, help in understanding toxicity responses by exploring connections between drugs or targets and related genes or chemicals. Users can also edit and expand networks based on the molecular relationships most relevant to the project.

IPA is capable of identifying pathways, molecular mechanisms and biological processes that are relevant to a given dataset. It is also capable of finding biological and chemical knowledge from the scientific literature. Other features allow for collaboration, sharing of results and insights with project teams.

IPA is a subscription-based software application. It is made available as a Web-based, hosted or deployed solution.

4.7 Gephi

Gephi is an open-source data exploratory, network visualization and analysis software tool for large network graphs. Gephi allows users to explore, analyze, spatialize, filter, cluster, manipulate, and export all types of network graphs. With Gephi, users can derive hypotheses and identify patterns by analyzing data using networks.

Gephi can be used to analyze a variety of networks ranging from biological networks to social networks. It supports the majority of the network file formats discussed in Subheading 2.2 above. The core of Gephi can perform basic network metric analysis such as calculating betweenness centrality, closeness, clustering, community detection or module identification. Gephi further includes a feature that allows for the analysis of dynamic networks where a set of networks representing or derived from different conditions or events are compared to infer differences. In addition, Gephi is also extensible by a range of plugins which users can install to perform functionality that is not included in the core of Gephi. While Gephi provides a range of network analysis features, other biological specific network analysis features such as functional enrichment cannot be easily done due to the unavailability of such functionality within Gephi or its associated plugins.

4.8 NDEx-The Network Data Exchange

NDEx-The Network Data Exchange is not so much a network analysis tool, but rather an open source framework for sharing of networks of many types and formats, publication of networks as data, and the use of networks in modular software [36]. Unlike other similar tools such as KEGG and IntAct, NDEx is a data commons framework that allows users to manage the sharing and the publication of networks. Users can upload any type of networks such as pathway models, interaction maps, and novel data-driven knowledge networks. NDEx supports networks of varying formats including simple interaction format (SIF), extensible graph markup and modeling language (XGMML), BioPAX3, and OpenBEL. Each network uploaded to NDEx is given an accession number which acts as a universally unique identifier allowing users to share or include such networks in publications. NDEx also promotes the development of network analysis algorithms and applications by providing access to networks which can be used as inputs through a Web-based relational state transfer application programming interface (REST API). In addition, users can anonymously access networks by searching through the Web interface (www.ndexbio.org). The framework can also be downloaded and run on a local server or personal computer, depending on the needs of a user.

4.9 PINA: Protein Interaction Analysis

Protein Interaction Analysis is a Web-based integrated network analysis platform for protein interaction network construction, filtering, analysis, visualization, and management [37]. PINA has a quarterly updated backend database consisting of an integration of data from six other publicly available databases; IntAct, MINT, BioGRID, DIP, HPRD, and MIPS MPact. To construct a network, PINA provides a query feature where users can either query a single protein, a list of proteins, a list of protein pairs or two lists of proteins.

The constructed PPI networks can be further analyzed by PINA's inbuilt GO term and protein domain annotation tools. Other analyses that can be performed include the use of graph theoretical tools to either discover basic topology properties of a PPI network or identify topologically important proteins, such as hubs or bottlenecks, based on several centrality measures from protein domains and GO terms. In addition, the constructed networks can be downloaded in customized tab-delimited, GraphML or MITAB formats for further analysis using tools such as Cytoscape where they can be integrated with gene expression profiles.

4.10 Colorectal Cancer Atlas

Colorectal Cancer Atlas [38] is an integrated Web-based resource mainly meant for those involved in colorectal cancer research. The tool provides a platform that catalogs both non-quantitative and quantitative proteomic and genomic sequence variation data in both colorectal cancer cell lines and tissues. This information has been curated from existing literature.

Colorectal Cancer Atlas features an easy to use search functionality that also offers auto-complete. Users can search for a given protein, gene, pathway, or cell line that may be of interest to them. Depending the type of search term, the tool then performs functional, pathway, and GO enrichment, maps sequence variances known in colorectal cancer and associated with the searched term, and generates a protein–protein interaction network.

The network integrates proteomic data with genomic sequence variations. Users can use this network analysis module to quickly get an overall picture of the interacting partners of a given gene in colorectal cancer. It uses color intensities to indicate the number of sequence variances for a given gene in the database. Users can also filter through the network by either a gene symbol or by cell lines.

While this tool is specific to colorectal cancer, it provides features that users can quickly use to get functional enrichment information for a given protein or gene as well as perform a gene or protein centered network analysis. Overall, researchers can quickly look up a list of genes or proteins and get an overview of a given gene in colorectal cancer.

4.11 Osprey

Osprey [39] is a software tool that allows for the visualization and analysis of complex interaction networks. Just like most visualization tools, in osprey genes are represented as nodes and interactions as edges. Developed using Java, Osprey is platform independent running on both Linux and Windows based systems.

Osprey provides a range of features that allows users to easily build data-rich graphical representations of their datasets. In addition, users can use the default BioGRID's Gene Ontology interaction datasets to quickly build an interaction network. Some of the features in Osprey include ability to compare functions between datasets, use of custom datasets to build interaction networks, ability to search for specific genes within a network as well filter functions to filter for specific nodes within a large a network. Osprey also has a number of network layouts including concentric circles, spoke, circular, and dual ring, these layouts allow for the comparison of large-scale datasets in an additive manner.

5 Conclusions

In order to study and understand complex systems such as cellular systems, we show that network theory provides metrics that can be used to study such systems using a bottom-up approach. In this chapter, we give an overview of how network theory can be applied to the analysis and study of proteomics data based on a number of network theory metrics. Such metrics include node degree, node centrality, Eigen vector values, and modularity.

We also discuss the most frequently used network analysis tools in analyzing proteomic data. In doing so, a generic workflow that one can use during the analysis is described. Tools discussed include databases which are used to house protein–protein interaction network annotations and the analytical tools that can be applied in analyzing proteomic data.

References

1. Mathivanan S (2014) Integrated bioinformatics analysis of the publicly available protein data shows evidence for 96% of the human proteome. J Proteomics Bioinformatics 2014(7):041–049. doi:10.4172/jpb.1000301
2. Kim M-S, Pinto SM, Getnet D, Nirujogi RS, Manda SS, Chaerkady R, Madugundu AK, Kelkar DS, Isserlin R, Jain S, Thomas JK, Muthusamy B, Leal-Rojas P, Kumar P, Sahasrabuddhe NA, Balakrishnan L, Advani J, George B, Renuse S, Selvan LDN, Patil AH, Nanjappa V, Radhakrishnan A, Prasad S, Subbannayya T, Raju R, Kumar M, Sreenivasamurthy SK, Marimuthu A, Sathe GJ, Chavan S, Datta KK, Subbannayya Y, Sahu A, Yelamanchi SD, Jayaram S, Rajagopalan P, Sharma J, Murthy KR, Syed N, Goel R, Khan AA, Ahmad S, Dey G, Mudgal K, Chatterjee A, Huang T-C, Zhong J, Wu X, Shaw PG, Freed D, Zahari MS, Mukherjee KK, Shankar S, Mahadevan A, Lam H, Mitchell CJ, Shankar SK, Satishchandra P, Schroeder JT, Sirdeshmukh R, Maitra A, Leach SD, Drake CG, Halushka MK, Prasad TSK, Hruban RH, Kerr CL, Bader GD, Iacobuzio-Donahue CA, Gowda H, Pandey A (2014) A draft map of the human proteome. Nature 509(7502):575–581. doi:10.1038/nature13302
3. Wilhelm M, Schlegl J, Hahne H, Gholami AM, Lieberenz M, Savitski MM, Ziegler E, Butzmann L, Gessulat S, Marx H, Mathieson T, Lemeer S, Schnatbaum K, Reimer U, Wenschuh H, Mollenhauer M, Slotta-Huspenina J, Boese J-H, Bantscheff M, Gerstmair A, Faerber F, Kuster B (2014) Mass-spectrometry-based draft of the human proteome. Nature 509(7502):582–587. doi:10.1038/nature13319
4. Pavlopoulos GA, Secrier M, Moschopoulos CN, Soldatos TG, Kossida S, Aerts J, Schneider R, Bagos PG (2011) Using graph theory to analyze biological networks. BioData Mining 4(1):1–27. doi:10.1186/1756-0381-4-10
5. Sevimoglu T, Arga KY (2014) The role of protein interaction networks in systems biomedicine. Comput Struct Biotechnol J 11(18):22–27. doi:10.1016/j.csbj.2014.08.008
6. Gandhi TKB, Zhong J, Mathivanan S, Karthick L, Chandrika KN, Mohan SS, Sharma S, Pinkert S, Nagaraju S, Periaswamy B, Mishra G, Nandakumar K, Shen B, Deshpande N, Nayak R, Sarker M, Boeke JD, Parmigiani G, Schultz J, Bader JS, Pandey A (2006) Analysis of the human protein interactome and comparison with yeast, worm and fly interaction datasets. Nat Genet 38(3):285–293, http://www.nature.com/ng/journal/v38/n3/suppinfo/ng1747_S1.html
7. Mathivanan S, Periaswamy B, Gandhi T, Kandasamy K, Suresh S, Mohmood R, Ramachandra Y, Pandey A (2006) An evaluation of human protein-protein interaction data in the public domain. BMC Bioinformatics 7(5):1–14. doi:10.1186/1471-2105-7-s5-s19
8. Pathan M, Keerthikumar S, Ang C-S, Gangoda L, Quek CYJ, Williamson NA, Mouradov D, Sieber OM, Simpson RJ, Salim A, Bacic A, Hill AF, Stroud DA, Ryan MT, Agbinya JI, Mariadason JM, Burgess AW, Mathivanan S (2015) FunRich: an open access standalone functional enrichment and interaction network analysis tool. Proteomics 15(15):2597–2601. doi:10.1002/pmic.201400515
9. Wilkins MR, Pasquali C, Appel RD, Ou K, Golaz O, Sanchez J-C, Yan JX, Gooley AA, Hughes G, Humphery-Smith I, Williams KL, Hochstrasser DF (1996) From proteins to proteomes: large scale protein identification by two-dimensional electrophoresis and amino acid analysis. Nat Biotechnol 14(1):61–65
10. Schmidt A, Forne I, Imhof A (2014) Bioinformatic analysis of proteomics data. BMC Syst Biol 8(Suppl 2):S3. doi:10.1186/1752-0509-8-S2-S3
11. Blais A, Dynlacht BD (2005) Constructing transcriptional regulatory networks. Genes Dev 19(13):1499–1511
12. De Las RJ, Fontanillo C (2010) Protein–protein interactions essentials: key concepts to building and analyzing interactome networks. PLoS Comput Biol 6(6), e1000807. doi:10.1371/journal.pcbi.1000807
13. Keshava Prasad TS, Goel R, Kandasamy K, Keerthikumar S, Kumar S, Mathivanan S, Telikicherla D, Raju R, Shafreen B, Venugopal A, Balakrishnan L, Marimuthu A, Banerjee S, Somanathan DS, Sebastian A, Rani S, Ray S, Harrys Kishore CJ, Kanth S, Ahmed M, Kashyap MK, Mohmood R, Ramachandra YL, Krishna V, Rahiman BA, Mohan S, Ranganathan P, Ramabadran S, Chaerkady R, Pandey A (2009) Human protein reference database—2009 update. Nucleic Acids Res 37(Database issue):D767–D772. doi:10.1093/nar/gkn892
14. Licata L, Briganti L, Peluso D, Perfetto L, Iannuccelli M, Galeota E, Sacco F, Palma A, Nardozza AP, Santonico E, Castagnoli L, Cesareni G (2012) MINT, the molecular interaction database: 2012 update. Nucleic Acids Res 40(D1):D857–D861. doi:10.1093/nar/gkr930
15. Chatr-Aryamontri A, Breitkreutz BJ, Oughtred R, Boucher L, Heinicke S, Chen D, Stark C,

Breitkreutz A, Kolas N, O'Donnell L, Reguly T, Nixon J, Ramage L, Winter A, Sellam A, Chang C, Hirschman J, Theesfeld C, Rust J, Livstone MS, Dolinski K, Tyers M (2015) The BioGRID interaction database: 2015 update. Nucleic Acids Res 43(Database issue):D470–D478. doi:10.1093/nar/gku1204

16. Szklarczyk D, Franceschini A, Wyder S, Forslund K, Heller D, Huerta-Cepas J, Simonovic M, Roth A, Santos A, Tsafou KP, Kuhn M, Bork P, Jensen LJ, von Mering C (2015) STRING v10: protein-protein interaction networks, integrated over the tree of life. Nucleic Acids Res 43(Database issue):D447–D452. doi:10.1093/nar/gku1003
17. Salwinski L, Miller CS, Smith AJ, Pettit FK, Bowie JU, Eisenberg D (2004) The database of interacting proteins: 2004 update. Nucleic Acids Res 32(suppl 1):D449–D451. doi:10.1093/nar/gkh086
18. Bader GD, Betel D, Hogue CW (2003) BIND: the biomolecular interaction network database. Nucleic Acids Res 31(1):248–250
19. Chatr-aryamontri A, Ceol A, Palazzi LM, Nardelli G, Schneider MV, Castagnoli L, Cesareni G (2007) MINT: the molecular INTeraction database. Nucleic Acids Res 35(suppl 1):D572–D574. doi:10.1093/nar/gkl950
20. Orchard S, Ammari M, Aranda B, Breuza L, Briganti L, Broackes-Carter F, Campbell NH, Chavali G, Chen C, del Toro N, Duesbury M, Dumousseau M, Galeota E, Hinz U, Iannuccelli M, Jagannathan S, Jimenez R, Khadake J, Lagreid A, Licata L, Lovering RC, Meldal B, Melidoni AN, Milagros M, Peluso D, Perfetto L, Porras P, Raghunath A, Ricard-Blum S, Roechert B, Stutz A, Tognolli M, van Roey K, Cesareni G, Hermjakob H (2014) The MIntAct project—IntAct as a common curation platform for 11 molecular interaction databases. Nucleic Acids Res 42(D1):D358–D363. doi:10.1093/nar/gkt1115
21. Shannon P, Markiel A, Ozier O, Baliga NS, Wang JT, Ramage D, Amin N, Schwikowski B, Ideker T (2003) Cytoscape: a software environment for integrated models of biomolecular interaction networks. Genome Res 13(11):2498–2504
22. Kohl M, Wiese S, Warscheid B (2011) Cytoscape: software for visualization and analysis of biological networks. Methods Mol Biol 696:291–303
23. Barabasi A-L, Oltvai ZN (2004) Network biology: understanding the cell's functional organization. Nat Rev Genet 5(2):101–113
24. Ho Y, Gruhler A, Heilbut A, Bader GD, Moore L, Adams S-L, Millar A, Taylor P, Bennett K, Boutilier K, Yang L, Wolting C, Donaldson I, Schandorff S, Shewnarane J, Vo M, Taggart J, Goudreault M, Muskat B, Alfarano C, Dewar D, Lin Z, Michalickova K, Willems AR, Sassi H, Nielsen PA, Rasmussen KJ, Andersen JR, Johansen LE, Hansen LH, Jespersen H, Podtelejnikov A, Nielsen E, Crawford J, Poulsen V, Sorensen BD, Matthiesen J, Hendrickson RC, Gleeson F, Pawson T, Moran MF, Durocher D, Mann M, Hogue CWV, Figeys D, Tyers M (2002) Systematic identification of protein complexes in Saccharomyces cerevisiae by mass spectrometry. Nature 415(6868):180–183
25. Ge H (2000) UPA, a universal protein array system for quantitative detection of protein–protein, protein–DNA, protein–RNA and protein–ligand interactions. Nucleic Acids Res 28(2):e3
26. Keene JD, Komisarow JM, Friedersdorf MB (2006) RIP-Chip: the isolation and identification of mRNAs, microRNAs and protein components of ribonucleoprotein complexes from cell extracts. Nat Protoc 1(1):302–307
27. Uetz P, Giot L, Cagney G, Mansfield TA, Judson RS, Knight JR, Lockshon D, Narayan V, Srinivasan M, Pochart P, Qureshi-Emili A, Li Y, Godwin B, Conover D, Kalbfleisch T, Vijayadamodar G, Yang M, Johnston M, Fields S, Rothberg JM (2000) A comprehensive analysis of protein-protein interactions in Saccharomyces cerevisiae. Nature 403(6770):623–627
28. Zahiri J, Bozorgmehr JH, Masoudi-Nejad A (2013) Computational prediction of protein–protein interaction networks: algorithms and resources. Curr Genomics 14(6):397–414. doi:10.2174/1389202911314060004
29. Pan A, Lahiri C, Rajendiran A, Shanmugham B (2015) Computational analysis of protein interaction networks for infectious diseases. Brief Bioinform. doi:10.1093/bib/bbv059
30. Jeong H, Mason SP, Barabasi AL, Oltvai ZN (2001) Lethality and centrality in protein networks. Nature 411(6833):41–42
31. Hartwell LH, Hopfield JJ, Leibler S, Murray AW (1999) From molecular to modular cell biology. Nature 402(6761 Suppl):47–52
32. Berenstein AJ, Piñero J, Furlong LI, Chernomoretz A (2015) Mining the modular structure of protein interaction networks. PLoS One 10(4), e0122477. doi:10.1371/journal.pone.0122477
33. Saito R, Smoot ME, Ono K, Ruscheinski J, Wang P-L, Lotia S, Pico AR, Bader GD, Ideker T (2012) A travel guide to Cytoscape plugins. Nat Methods 9(11):1069–1076. doi:10.1038/nmeth.2212

34. Han K, Park B, Kim H, Hong J, Park J (2004) HPID: The human protein interaction database. Bioinformatics 20(15):2466–2470. doi:10.1093/bioinformatics/bth253
35. Chen JY, Mamidipalli S, Huan T (2009) HAPPI: an online database of comprehensive human annotated and predicted protein interactions. BMC Genomics 10(Suppl 1):S16
36. Pratt D, Chen J, Welker D, Rivas R, Pillich R, Rynkov V, Ono K, Miello C, Hicks L, Szalma S, Stojmirovic A, Dobrin R, Braxenthaler M, Kuentzer J, Demchak B, Ideker T (2015) NDEx, the network data exchange. Cell Syst 1(4):302–305. doi:10.1016/j.cels.2015.10.001
37. Cowley MJ, Pinese M, Kassahn KS, Waddell N, Pearson JV, Grimmond SM, Biankin AV, Hautaniemi S, Wu J (2012) PINA v2.0: mining interactome modules. Nucleic Acids Res 40(D1):D862–D865. doi:10.1093/nar/gkr967
38. Chisanga D, Keerthikumar S, Pathan M, Ariyaratne D, Kalra H, Boukouris S, Mathew NA, Saffar HA, Gangoda L, Ang C-S, Sieber OM, Mariadason JM, Dasgupta R, Chilamkurti N, Mathivanan S (2016) Colorectal cancer atlas: an integrative resource for genomic and proteomic annotations from colorectal cancer cell lines and tissues. Nucleic Acids Res 44(D1):D969–D974. doi:10.1093/nar/gkv1097
39. Breitkreutz B-J, Stark C, Tyers M (2003) Osprey: a network visualization system. Genome Biol 4(3):R22

Chapter 15

Determining the Significance of Protein Network Features and Attributes Using Permutation Testing

Joseph Cursons and Melissa J. Davis

Abstract

Network analysis methods are increasing in popularity. An approach commonly applied to analyze proteomics data involves the use of protein–protein interaction (PPI) networks to explore the systems-level cooperation between proteins identified in a study. In this context, protein interaction networks can be used alongside the statistical analysis of proteomics data and traditional functional enrichment or pathway enrichment analyses. In network analysis it is possible to adjust for some of the complexities that arise due to the known, explicit interdependence between the measured quantities, in particular, differences in the number of interactions between proteins. Here we describe a method for calculating robust empirical *p*-values for protein interaction networks. We also provide a worked example with python code demonstrating the implementation of this methodology.

Key words PPI, Protein interaction, Network structure, Permutation testing, Computational systems biology, PROSPERITI, Proteomics

1 Introduction

Network analysis using protein–protein interactions to interpret biological data has become popular in recent years as researchers become increasingly interested in the identification of emergent, systems-level effects in their experimental models [1–4]. Protein interaction network analysis is particularly appropriate for proteomics data, as the networks relate directly to the molecules being measured [5]. A common approach for the network analysis of proteomic data is to map proteins detected in an experiment, or varying in abundance between two conditions, onto a network of known protein–protein interactions. The resulting network is then explored for features thought to be associated with the biological question motivating the study. Network features of interest often include topological features (metrics describing connectivity and size), the existence of modules/motifs (functionally significant sub-networks) or hubs (highly connected nodes), or enrichment with Gene Ontology terms of interest.

Shivakumar Keerthikumar and Suresh Mathivanan (eds.), *Proteome Bioinformatics*, Methods in Molecular Biology, vol. 1549, DOI 10.1007/978-1-4939-6740-7_15,

What these approaches often lack however is a principled way to assess the statistical significance for features of interest. Often researchers will compare their derived network to randomly generated networks [6] to demonstrate that their network is significantly different to random networks. It has been shown that virtually all biological networks share properties such as scale free degree distributions and short diameters, and we assert that it is not meaningful to search for differences between a network of interest and random networks with fundamentally different characteristics.

In other cases, common statistical tests are applied to determine if a network is enriched for a property of interest when compared to a background set. Development of these methods has been grounded in approaches that determine functional enrichment from high-throughput expression study gene lists, and they effectively discard connectivity in the underlying biological network before performing statistical tests on node sets extracted from this network [7]. The use of statistical tests in this fashion has serious limitations; in particular, the structure of the 'true biological' network that proteins operate within can influence quantitative measures leading to spurious conclusions. Underlying assumptions around the independence of variable measurements are fundamentally flawed in scenarios where proteins (or genes) are selected based on their relationships within a network. Permutation testing represents an attractive alternative to the application of statistical tests. A number of excellent resources [8–10] describe the statistical theories that underlie permutation testing. Here, we focus on a practical implementation of this strategy for network analysis, and describe a methodology to determine the statistical significance of network features (given a well-articulated hypothesis) through the use of permutation testing to generate an empirical null distribution [8].

This work follows a standard methodology for network construction and analysis, where we start with a list of proteins that arise from an experiment, then build a network from these proteins using a public knowledge base of known, experimentally defined protein interactions. Then, we apply permutation testing to see if any network features (topological metrics, or statistical associations within the data) are significant in the context of that network. A worked example of this approach is provided as a python script, using a published set of tyrosine phosphorylation data generated from breast cancer cell lines [11], and a publicly available protein interaction network [12]. Readers are encouraged to download this code and work through the graphical README in conjunction with this chapter.

In the study from which we derive our phospho-protein abundance data the researchers highlight a densely connected phosphoprotein signaling network for basal breast cancer cell lines which is centered around the Src family kinase member Lyn [11]. Here, we integrate protein–protein interactions from PINA to quantify the size (measured by diameter and the number of connected notes) and density (average clustering coefficient) of the phospho-protein network identified across all cell lines (the 'background network') and for the MDA-MB-231 cell line (the 'condition specific network'). We then use permutation testing to estimate the statistical significance of these quantitative parameters. As shown below, the diameter of the background network is within the expected range; however, the diameter of the MDA-MB-231 network is equivalent to the background network and this is greater than what would be expected given the number of unique proteins measured. We also show that the average clustering coefficient and number of connected nodes is much greater than expected for both the background and MDA-MB-231 network. These results likely reflect the high-degree of "inter-connectivity" within the phospho-tyrosine signaling network. Furthermore, analysis of the MDA-MB-231 network provides quantitative support for the observation [11] that this basal breast cancers appear to contain a "prominent SFK [Src family kinase] signalling network."

2 Materials

2.1 Computational Scripts

A python script containing example code for performing this analysis can be downloaded from the GitHub Project—**Pro**tein network **s**ignificance **per**mutat**i**on **t**est**i**ng: http://github.com/DavisLaboratory/PROSPERITI.

A graphical README on the GitHub page contains cross-references between this chapter and the corresponding computational script.

2.2 Data

This analysis uses phospho-tyrosine enriched protein measurements across breast cancer cell lines [11], in particular Supplemental Table 3, which can be downloaded at: http://cancerres.aacrjournals.org/content/70/22/9391/suppl/DC1.

Protein–protein interaction data from the Protein Interaction Network Analysis (PINA) Platform v2.0 [12] were downloaded in the MI-TAB format. MI-TAB is a standard format for representing protein interaction data that uses the Protein Standards Initiative ontology for molecular interactions [13]. These data can be downloaded directly at: http://cbg.garvan.unsw.edu.au/pina/download/Homo%20sapiens-20140521.tsv.

3 Methods

As noted above, there is a graphical README on the GitHub project page for PROSPERITI, which cross-references these methods to the corresponding computational script.

3.1 Standard Network Construction

1. Collect proteomics data and identify the results of interest; for some, this result may only involve proteins with large changes in abundance; here we consider the detection of a phospho-tyrosine peptide to be indicative of active signaling for that protein, and thus we examine all proteins identified in a particular experiment. The phospho-tyrosine enriched MS/MS data examined here contain measurements for 303 nonunique proteins (265 unique proteins; due to peptide identity) over 15 different breast cancer cell lines (Fig. 3a).
2. If necessary, convert proteins of interest to UniProt accession numbers, a standard identifier for protein data used across protein interaction databases. Resources such as the Ensembl BioMart (http://www.ensembl.org/info/data/biomart/index.html) provide identifier conversion services if required. In our example, UniProt identifiers are provided (Fig. 1a) and we use the full list, detected across all cell lines, to construct the background network. Condition specific lists can also be constructed from the individual cell lines.
3. Build a protein interaction network by identifying known interactions between selected proteins — here we use a comprehensive list of protein–protein interactions from PINA v2.0 (Fig. 1b). Protein interaction networks can be constructed so they capture interactions only between proteins in the results (a zeroth order network), or to capture interactions between proteins in the results and other proteins (a first order network) (*see* Fig. 2). Wider networks can be constructed, but at greater than two steps from a given protein, networks can become very large.

3.2 Defining a Null Hypothesis and Designing the Permutation Testing

Calculation of empirical *p*-values through permutation testing requires an explicitly stated hypothesis in order to determine the most appropriate way to model the distribution of metrics under the null hypothesis. Care should be taken to identify the hypothesis to be tested, or p-values, no matter how calculated, can be misleading or meaningless.

Here, we are going to test two common hypotheses (*see* **Note 2**, Section 4 for other examples):

1. That a protein connectivity metric (e.g., describing connectivity) is significant. To consider protein connectivity metrics with confidence, the observed value must be compared to the distributions of connectivity metrics that are generated under the null hypothesis. In this case, the null hypothesis states that any randomly selected set of proteins (of the same size as the result-generating network; Fig. 1d) will be able to generate a network with similar topological features (and thus with similar connectivity metrics; Fig. 1f).
2. That interacting proteins have correlated (phospho-)protein abundance. Here we hypothesize that all protein pairs selected

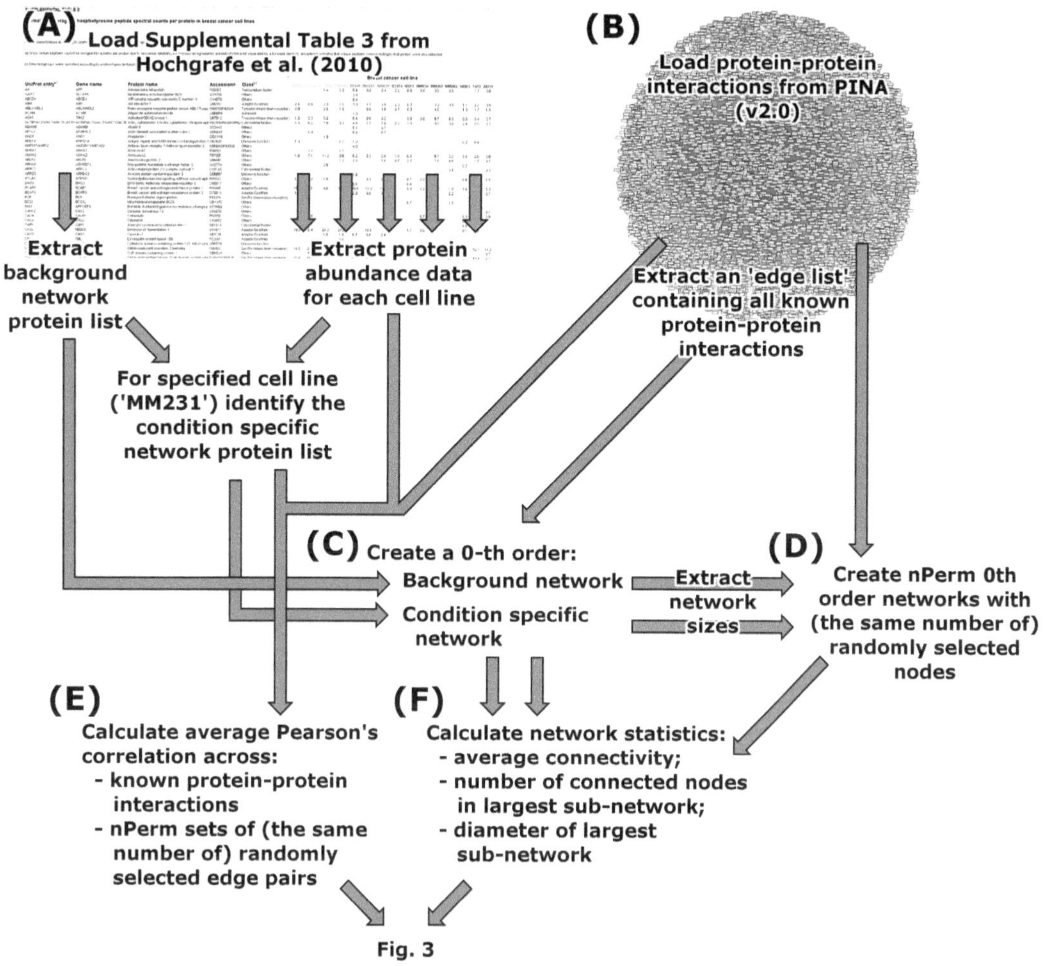

Fig. 1 Graphical workflow for network construction and analysis. (**a**) Identify proteins from the experiment of interest—here, we extract UniProt identifiers and relative phospho-protein abundance data from a published report [11]. All UniProt identifiers (*i.e.* across all cell lines) form the background network list, while individual cell lines produce condition specific lists—in our example we examine the MDA-MB-231 (MM231) cell line. (**b**) Load in the full set of protein–protein interactions from PINA (v2.0) [12]. (**c**) Construct zeroth order interaction networks for the background and condition specific data, using nodes (proteins) from (**a**) and edges (interactions) from (**b**). (**d**) Apply permutation testing to create the specified number of random networks (n_{Perm}, here we use 10,000) with the same number of starting nodes. (**e**) Using the full set of phospho-protein abundance data from (**a**) and the protein–protein interaction list from (**b**), calculate the network-wide average absolute correlation. A background distribution was generated by measuring the average, absolute correlation across the same number of randomly selected edges, while excluding known protein–protein interactions. Results are shown in Fig. 3. (**f**) Calculate properties of interest for the networks from (**c**), and estimate the background distribution using the networks from (**d**). In this example we measure: the number of nodes within the largest connected sub-network; the diameter of the largest connected sub-network; and the average clustering coefficient across all nodes. Results are shown in Fig. 3

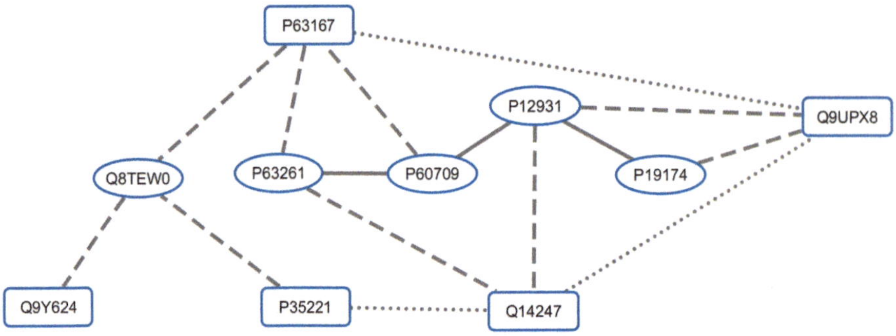

Fig. 2 Different levels of network expansion. Zeroth order, first order and second order networks illustrating the expansion seen at each step in network construction. A set of five proteins (ellipses) are used to seed a network; zeroth order interactions between these proteins have solid lines, first order interactions have dashed lines, and draw a further five proteins (rectangles) into an extended network; second order interactions have dotted lines. Zeroth order networks are constructed by querying the interactome to identify interactions where both interacting partners are seed nodes; this is the type of network we construct in our worked example (Fig. 1c). First order networks are constructed by querying the full PPI to identify interactions involving at least one of the seed nodes

by their interactions will a show higher average absolute correlation (as a general and unidirectional measure of statistical association; Fig. 1f). The null hypothesis here is that randomly selected pairs of measured proteins would show a similar correlation (*i.e.* correlations observed between the data are not associated with the underlying protein–protein interactions).

3.2.1 Testing Network Topology

To test hypothesis 1 above:

1. Generate a random set of proteins equivalent in size to the network being tested (here 265 proteins for the background network and 79 proteins for the condition specific network).
2. Build a zeroth-order network from this random set of proteins using the method described in Section 3.1.
3. For topological descriptors of interest calculate appropriate metrics. Here, the average clustering coefficient is calculated, then, the largest connected sub-network is identified, allowing us to measure the diameter of, and the corresponding number of nodes within this sub-network.
4. Repeat (**steps 1–3**) 10,000 times to build up a distribution for each topological metric (Fig. 3a–c, e–f).
5. Compare the observed value of topological metrics from the experimental network (Fig. 3, *red vertical line*) to the null distribution generated above.
6. Determine if the topological descriptors are significant:
   - If the background distribution is approximately Normal, it may be easiest to calculate the *Z*-score (difference from the

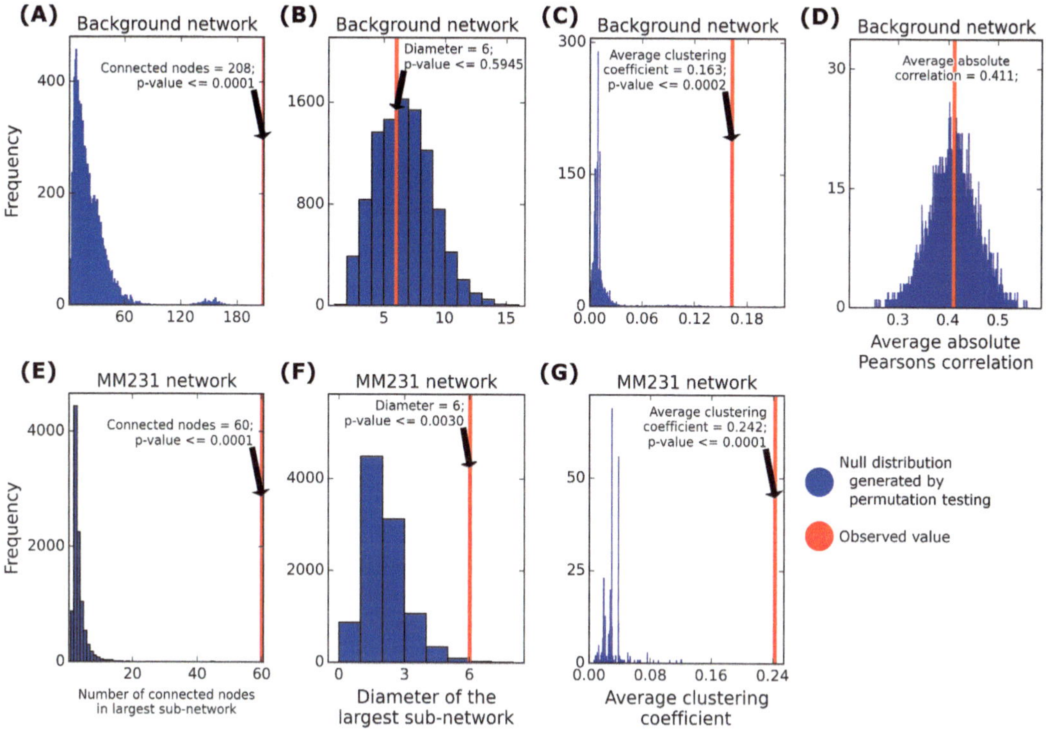

Fig. 3 Quantitative network features shown relative to null distributions generated using permutation testing. Quantitative network features calculated for (**a**–**d**) the background network and (**e**–**f**) the MDA-MB-231 (MM231) condition-specific network. For all plots, *blue histograms* show the background distribution generated using permutation testing, while the observed value is shown with a red vertical line. (**a**, **e**) The number of nodes within the largest connected sub-network. (**b**, **f**) The diameter of the largest connected sub-network. (**c**, **g**) The average clustering coefficient—note that the frequency axes have been scaled to exclude the first bin (average clustering coefficient = 0) which contains many observations. (**d**) The average absolute correlation (between phospho-protein abundance, where at least five matched observations are present; this cannot be calculated for the single MM231 sample)

sample mean, normalized by the sample standard deviation) and then convert to an empirical p-value.

– Alternatively, an empirical p-value can be estimated by examining the position of the observed value relative to the cumulative density function of the permutation test distribution. It should be noted that a sufficiently high value of n_{Perm} needs to be selected (*see* **Note 1** *in* Section 4).

3.2.2 Testing Association Between the Data and the Network

In this case, the network that is being sampled is the network of all pairwise correlations between proteins in the experiment. The experimental sample is based on known protein interactions between those proteins.

1. For all known protein–protein interactions within the background network where there are at least five matched observations, calculate the average absolute Pearson's correlation.

2. Randomly generate a set of edges from the pairwise correlation network equivalent in size to the number of edges in the known PPI network, while excluding known PPIs.
3. Calculate the average correlation of this network.
4. Repeat (**steps 2** and **3**) 10,000 times to build up a distribution of the average correlations (*see* Fig. 3d, *blue histograms*).
5. Compare the observed value of the average correlation in the experimental network to the distribution generated above (Fig. 3d, *red vertical line*).
6. Determine if the statistical association is significant as described in **Step 6** of Section 3.2.1.

4 Notes

1. An advantage of this strategy is that when comparing the experimentally determined value against an empirical null distribution, it is relatively easy to estimate the false discovery rate. This explicit modelling of false discovery likelihood assists clear interpretation, and largely determines the minimum number of permutations that should be performed [14], although a greater number of permutations generally gives better estimates of the null distribution and more robust estimates of significance.
2. Other common hypotheses not worked in our example are:
   1. That a network is enriched for a particular function.

      In this case, it is common to see standard Gene Ontology or Pathway enrichment tests applied. These tests often assume independence between GO terms or pathways, and this is often invalid in the context of a protein-protein interaction network. Permutation testing gives a robust estimate for the significance of GO terms or pathway annotation enrichment in the network. Here the method would be:

      (a) Generate 10,000 random protein lists.

      (b) Build 10,000 networks from these random lists.

      (c) Identify the number of proteins in each network that are part of your pathway or GO category of interest—this is the distribution of associations between networks and the term or pathway under the null hypothesis.

      (d) Compare the observed number of proteins in the experimentally derived network to this distribution and calculate an estimated p-value as discussed above (**Step 6** in Section 3.2.1).

2. The presence of a particular hub-node in a network is meaningful.

   In this case, it is important to build networks from random protein sets and identify how frequently a particular hub-node may emerge. Some proteins have so many interactions that they will frequently emerge as hubs.

   (a) Generate 10,000 random protein lists.

   (b) Build 10,000 networks from these random lists.

   (c) Calculate the number of times your protein of interest is detected as a hub in random networks; the best strategy for this is to plot the distribution of that protein's degree across the random networks (the degree will be 0 in networks where the protein is not present).

   (d) Compare the observed degree of your hub protein to this distribution and use the *Z*-score to calculate an estimated *p*-value as discussed above.

Protein–protein interaction networks can also be used to analyze coordinated behavior in proteins identified through mRNA transcript data where differential abundance is expected to disrupt the underlying signaling or protein–protein interaction networks [15]. Although care should be taken when interpreting the results of such analyses, as changes in transcript abundance are not always concordant with changes in protein abundance or activity, this can provide context to aid with interpreting results, and the procedures described here are also valid for determining significance in these applications.

Finally, these methods are appropriate for application to studies where networks are constructed from a known background network. Statistical analysis of putative regulatory networks inferred from data face different challenges which are not explored here. The code provided as a supplement to this chapter can be adapted to address these and other hypotheses of interest to researchers working in network analysis.

Acknowledgment

M.J.D. is supported by the Australian NBCF CDF-14-043. J.C. is funded by the Australian Research Council Centre of Excellence in Convergent Bio-Nano Science and Technology (project number CE140100036). We thank colleagues within the Systems Biology Laboratory at the University of Melbourne and the Bioinformatics Division at the Walter and Eliza Hall Institute for constructive suggestions and useful discussions around this topic.

References

1. Aderem A (2005) Systems biology: its practice and challenges. Cell 121(4):511–513. doi:10.1016/j.cell.2005.04.020
2. Genovesi LA, Ng CG, Davis MJ, Remke M, Taylor MD, Adams DJ, Rust AG, Ward JM, Ban KH, Jenkins NA, Copeland NG, Wainwright BJ (2013) Sleeping beauty mutagenesis in a mouse medulloblastoma model defines networks that discriminate between human molecular subgroups. Proc Natl Acad Sci U S A 110(46):E4325–4334. doi:10.1073/pnas.1318639110
3. Gajadhar AS, White FM (2014) System level dynamics of post-translational modifications. Curr Opin Biotechnol 28:83–87. doi:10.1016/j.copbio.2013.12.009
4. Sevimoglu T, Arga KY (2014) The role of protein interaction networks in systems biomedicine. Comput Struct Biotechnol J 11(18):22–27. doi:10.1016/j.csbj.2014.08.008
5. Inder KL, Davis M, Hill MM (2013) Ripples in the pond—using a systems approach to decipher the cellular functions of membrane microdomains. Mol Biosyst 9(3):330–338. doi:10.1039/c2mb25300c
6. Sun J, Zhao Z (2010) A comparative study of cancer proteins in the human protein-protein interaction network. BMC genomics 11(Suppl 3):S5. doi:10.1186/1471-2164-11-S3-S5
7. Khatri P, Sirota M, Butte AJ (2012) Ten years of pathway analysis: current approaches and outstanding challenges. PLoS Comput Biol 8(2):e1002375. doi:10.1371/journal.pcbi.1002375
8. Phipson B, Smyth GK (2010) Permutation P-values should never be zero: calculating exact P-values when permutations are randomly drawn. Stat Appl Genet Mol Biol 9:39. doi:10.2202/1544-6115.1585
9. Ernst MD (2004) Permutation methods: a basis for exact inference. Stat Sci 19(4):676–685
10. Good P (2013) Permutation tests: a practical guide to resampling methods for testing hypotheses. Springer Science & Business Media, New York, NY
11. Hochgrafe F, Zhang L, O'Toole SA, Browne BC, Pinese M, Porta Cubas A, Lehrbach GM, Croucher DR, Rickwood D, Boulghourjian A, Shearer R, Nair R, Swarbrick A, Faratian D, Mullen P, Harrison DJ, Biankin AV, Sutherland RL, Raftery MJ, Daly RJ (2010) Tyrosine phosphorylation profiling reveals the signaling network characteristics of basal breast cancer cells. Cancer Res 70(22):9391–9401. doi:10.1158/0008-5472.CAN-10-0911
12. Cowley MJ, Pinese M, Kassahn KS, Waddell N, Pearson JV, Grimmond SM, Biankin AV, Hautaniemi S, Wu J (2012) PINA v2.0: mining interactome modules. Nucleic Acids Res 40(Database Issue):D862–865. doi:10.1093/nar/gkr967
13. Hermjakob H (2006) The HUPO proteomics standards initiative—overcoming the fragmentation of proteomics data. Proteomics 6(Suppl 2):34–38. doi:10.1002/pmic.200600537
14. Knijnenburg TA, Wessels LF, Reinders MJ, Shmulevich I (2009) Fewer permutations, more accurate P-values. Bioinformatics 25(12):i161–168. doi:10.1093/bioinformatics/btp211
15. Cursons J, Leuchowius KJ, Waltham M, Tomaskovic-Crook E, Foroutan M, Bracken CP, Redfern A, Crampin EJ, Street I, Davis MJ, Thompson EW (2015) Stimulus-dependent differences in signalling regulate epithelial-mesenchymal plasticity and change the effects of drugs in breast cancer cell lines. Cell Commun Signal 13:26. doi:10.1186/s12964-015-0106-x

Chapter 16

Bioinformatics Tools and Resources for Analyzing Protein Structures

Jason J. Paxman and Begoña Heras

Abstract

The dramatic increase in the number of protein sequences and structures deposited in biological databases has led to the development of many bioinformatics tools and programs to manage, validate, compare, and interpret this large volume of data. In addition, powerful tools are being developed to use this sequence and structural data to facilitate protein classification and infer biological function of newly identified proteins. This chapter covers freely available bioinformatics resources on the World Wide Web that are commonly used for protein structure analysis.

Key words Protein Structure, Protein Data Bank (PDB), PDBe, PDBj, MolProbity, PDB-REDO, PDBsum, PDBePISA, DALI, PDBeMotif, ProFunc

1 Introduction

Understanding the function of proteins in cellular processes, infection and disease is fundamental to many disciplines of science and medicine. The amino acid sequence, protein structure, or the types of ligands that a protein binds is valuable information that can be used to better understand protein function. The advent of next generation sequencing has led to an exponential growth in the number of fully sequenced genomes, whereby there are currently over 8000 genomes in public databases [1]. This has resulted in the deposition of more than 60 million unique protein sequences in the UniProt Knowledgebase (UniProtKB [2]). Similarly, advances in crystallography such as increased automation of sample handling and diffraction data processing have resulted in an ever increasing number of protein structures submitted to online databases such as the Protein Data Bank (PDB) [3]. Currently over 120,000 protein structures (including protein–DNA and protein–ligand complexes) have been deposited in the PDB. Protein structures can offer more insight into biological function than sequences alone as they show the three-dimensional arrangement

Shivakumar Keerthikumar and Suresh Mathivanan (eds.), *Proteome Bioinformatics*, Methods in Molecular Biology, vol. 1549, DOI 10.1007/978-1-4939-6740-7_16, © Springer Science+Business Media LLC 2017

of interacting residues and the overall structures show higher conservation than the amino acid sequences [4]. Despite the increase in the availability of protein structures, this rate still lags behind the deposition of protein sequences.

The wealth of information together with the World Wide Web (WWW) has seen the continuous development and expansion of bioinformatics tools and resources to manage the growing sequence/structural data, ensure the quality of the structural data, facilitate its interpretation, and allow comparative analyses of protein structures and sequences to inform protein classification and biological function. Many of these online bioinformatics tools are designed for different levels of expertise, whereby even the most experienced scientist will find many of these resources a welcome addition to their tool kit.

This chapter summarizes a number of freely available bioinformatics tools and databases currently available on the WWW, primarily for the analysis of protein structures determined by X-ray crystallography.

2 Methods

2.1 Structure Validation is the First Step

Structural biologists aim at obtaining the most accurate model to describe their data, which is essential to draw conclusions about the function of the protein under study. Structure validation is therefore a critical starting point in the analysis of the three-dimensional structures of macromolecules.

2.1.1 Analyzing the Experimental Diffraction Data and Atomic Model

Diffraction data based validation parameters include the diffraction data resolution and the crystallographic *R*work and *R*free values, which measure the degree to which the macromolecular structures fit the experimental data and indicate a potential bias in the model building process [5]. Other validation parameters are the *B* factor which shows the confidence in the positioning of the different atoms and reflects the degree of order in the crystal. The real-space *R*-value (RSR) or the RSR-*Z* score (RSR normalized for specific residue type and resolution shell) [6] defines how the atomic model fits the experimental data in real space.

The quality of the atomic model is assessed by defining the clash-score, which measures steric clashes between pairs of atoms. Other quality indicators include the Ramachandran and side-chain analysis, which measure deviations of the backbone and side chain torsion angles from standard values, respectively. Structural validation tools such as PROCHECK [7] (https://www.ebi.ac.uk/thornton-srv/software/PROCHECK/), WHAT IF [8], and more recently MolProbity (http://molprobity.biochem.duke.edu) [9] (Table 1) are commonly used to calculate most of the validation parameters and evaluate the overall quality of three-dimensional

Table 1
Selection of protein structure validation tools

Server	URL	Description	Ref.
PROCHECK	https://www.ebi.ac.uk/thornton-srv/software/PROCHECK/	Structural validation tool	[7]
WHAT IF	http://swift.cmbi.ru.nl/servers/html/index.html	Structural validation tool	[8]
MolProbity	http://molprobity.biochem.duke.edu	Structure validation and scoring	[9]
wwPDB Validation Reports	http://wwpdb-validation.wwpdb.org/validservice	Quality assessment considering atomic model and diffraction data	[11]
PURY	http://pury.ijs.si/	Resource to asses the geometry of ligands in PDB files	[12]
Twilight	http://www.ruppweb.org/twilight/default.htm	Tool to analyze and correct ligand geometry in PDB files	[13]
pdb-care	http://www.glycosciences.de/tools/pdb-care/	Tool to analyze carbohydrate structures in PDB files	[14]
PDB_REDO	https://xtal.nki.nl/PDB_REDO/	Tool to diagnose and automatically re-refine protein crystal structures	[15]

structures of proteins, nucleic acids, glycoproteins, and protein–carbohydrate/nucleic acid complexes. MolProbity also includes a number of other programs for structure validation such as REDUCE which adds hydrogens to a molecular structure file, along with PROBE (evaluates close contacts between atoms), RAMALYZE (validates protein backbone Ramachandran dihedral angles) and ROTALYZE (validates protein side chain rotamers). Additionally, it includes programs such as DANGLE and SUITENAME, which assess the ideality of nucleic acid and protein backbone geometry [10].

2.1.2 Structure Validation: Scoring

MolProbity provides a "MolProbity score," which is an overall measure of the quality of the three dimensional protein structure (Fig. 1). This score is calculated combining the validation parameters clashscore, Ramachandran and rotamer outliers and compares these quality measures to other structures in the PDB of comparable resolution (e.g. a MolProbity score lower than the crystallographic resolution indicates that the structure quality is better than the average structure at a comparable resolution). Other recently developed validation tools include wwPDB Validation Reports

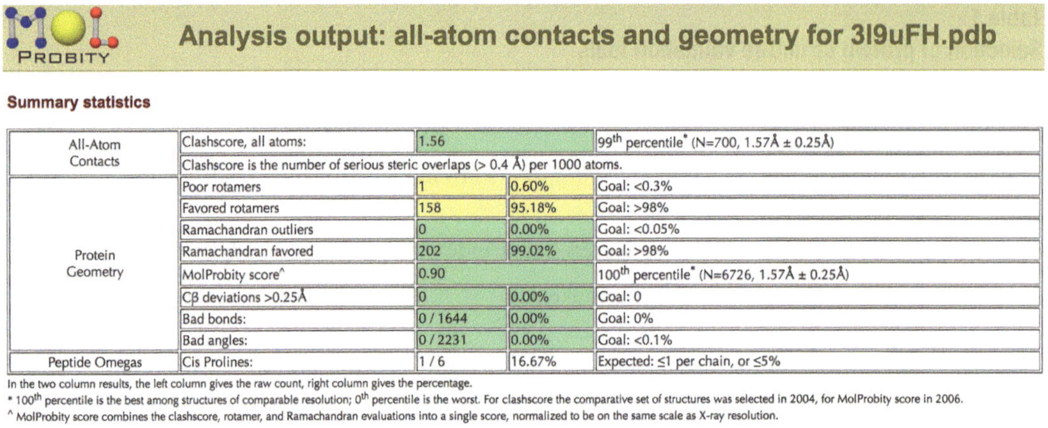

Analysis output: all-atom contacts and geometry for 3I9uFH.pdb

Summary statistics

All-Atom Contacts	Clashscore, all atoms:	1.56		99th percentile* (N=700, 1.57Å ± 0.25Å)
	Clashscore is the number of serious steric overlaps (> 0.4 Å) per 1000 atoms.			
Protein Geometry	Poor rotamers	1	0.60%	Goal: <0.3%
	Favored rotamers	158	95.18%	Goal: >98%
	Ramachandran outliers	0	0.00%	Goal: <0.05%
	Ramachandran favored	202	99.02%	Goal: >98%
	MolProbity score^	0.90		100th percentile* (N=6726, 1.57Å ± 0.25Å)
	Cβ deviations >0.25Å	0	0.00%	Goal: 0
	Bad bonds:	0 / 1644	0.00%	Goal: 0%
	Bad angles:	0 / 2231	0.00%	Goal: <0.1%
Peptide Omegas	Cis Prolines:	1 / 6	16.67%	Expected: ≤1 per chain, or ≤5%

In the two column results, the left column gives the raw count, right column gives the percentage.
* 100th percentile is the best among structures of comparable resolution; 0th percentile is the worst. For clashscore the comparative set of structures was selected in 2004, for MolProbity score in 2006.
^ MolProbity score combines the clashscore, rotamer, and Ramachandran evaluations into a single score, normalized to be on the same scale as X-ray resolution.

Fig. 1 A MolProbity results summary for the disulfide oxidoreductase TcpG at 1.2 Å resolution (PDB code: 4DVC [40]) which provides the overall quality of a macromolecular three-dimensional structure including values, goals and relative percentiles for all-atoms, clashscore and protein geometry criteria. It uses traffic light color-coding (*red/yellow/green*) to indicate poor to more favorable values

(http://wwpdb-validation.wwpdb.org/validservice/) (Table 1), which assess the quality of the structure taking into consideration the atomic model and the experimental diffraction data to provide an at-a-glance summary of the results [11].

2.1.3 Structure Validation: Protein–Ligand Complexes

Resources for analyzing protein–ligand complexes have also been developed. PURY [12] (http://pury.ijs.si/) assesses the geometry of ligands and creates topology and parameter files compatible with common refinement programs. Twilight [13] (http://www.ruppweb.org/twilight/default.htm) analyzes and corrects ligands that have low correlation with the corresponding electron density maps (Table 1). Another useful resource is pdb-care [14] (http://www.glycosciences.de/tools/pdb-care/) a tool developed to analyze carbohydrate structures in PDB files (Table 1).

2.1.4 Structure Validation: Parameter Optimization

Tools to diagnose and automatically re-refine protein crystal structures are also currently available. For example, the PDB_REDO [15] project was developed to re-refine old crystal structures in the Protein Data Bank (PDB) using the latest structure refinement and validation techniques. Through the PDB_REDO server (https://xtal.nki.nl/PDB_REDO/) (Table 1) the scientific community has access to all the re-refined structures. Furthermore, this web server also offers a tool for macromolecular X-ray crystallographers to automatically optimize the refinement of their protein structure. This resource does not rebuild the original model but involves automatic optimization of parameters such as TLS groups (predicts local displacement of atoms), X-ray and B-factor weights and rebuilding and flipping side-chains into favorable rotamer conformations with optimized hydrogen bonding networks.

2.2 Standard Tools for Analyzing Protein Structures

Several bioinformatics resources are available through the Protein Data Bank Europe (PDBe, http://www.ebi.ac.uk/pdbe), the Protein Data Bank Japan (PDBj, http://pdbj.org) and other databases for the analysis of protein structures.

2.2.1 Sequence and Structure Analysis

The web database PDBsum [16] (https://www.ebi.ac.uk/thornton-srv/databases/cgi-bin/pdbsum/GetPage.pl?pdbcode=index.html) (Table 2) provides a pictorial summary of multiple structural analyses for PDB entries or any PDB file uploaded by the user. The PDBsum summary (Fig. 2a) shows all the contents of any PDB (protein chains, nucleotide chains, ligands, and water molecules), the results from the PROCHECK quality assessment program and provides links for viewing the coordinates in three-dimensions using web-based visualization tools such as JMol (http://jmol.sourceforge.net/) or RasMol [17]. Additionally, PDBsum gives schematic diagrams illustrating a number of structural analyses. For example, the Pfam domain diagram displays all constituent Pfam domains [18], http://pfam.xfam.org/), the protein's secondary structure computed using PROMOTIF [19], and the CATH structural classification organization [20]. PDBsum also provides numerous links to related data in other databases. For example, all Pfam domains shown in the Pfam domain diagram are hyperlinked to other PDB entries containing those domains.

Table 2
Selection of protein structure analysis resources

Server	URL	Description	Ref.
PDBsum	https://www.ebi.ac.uk/thornton-srv/databases/cgi-bin/pdbsum/GetPage.pl?pdbcode=index.htm	Compilation of tools to perform multiple structural analyses	[16]
MAFFTash	http://sysimm.ifrec.osaka-u.ac.jp/MAFFTash/	Multiple sequence alignment tool using sequence and structural data	[22]
PDBePISA	http://www.ebi.ac.uk/msd-srv/prot_int/pistart.html	Tool to analyze protein interfaces and quaternary structure prediction	[24]
DrugPort	http://www.ebi.ac.uk/thornton-srv/databases/drugport/	Surveys the PDB for structural information related to a query drug molecule	[16]
PDBeXpress	http://www.ebi.ac.uk/pdbe-srv/pdbexpress/	Collection of tools to extract from the PDB protein–ligand interaction statistics	
eF-seek	http://pdbj.org/help/ef-seek	Tool to examine the PDB for proteins with similar ligand binding sites	[27]

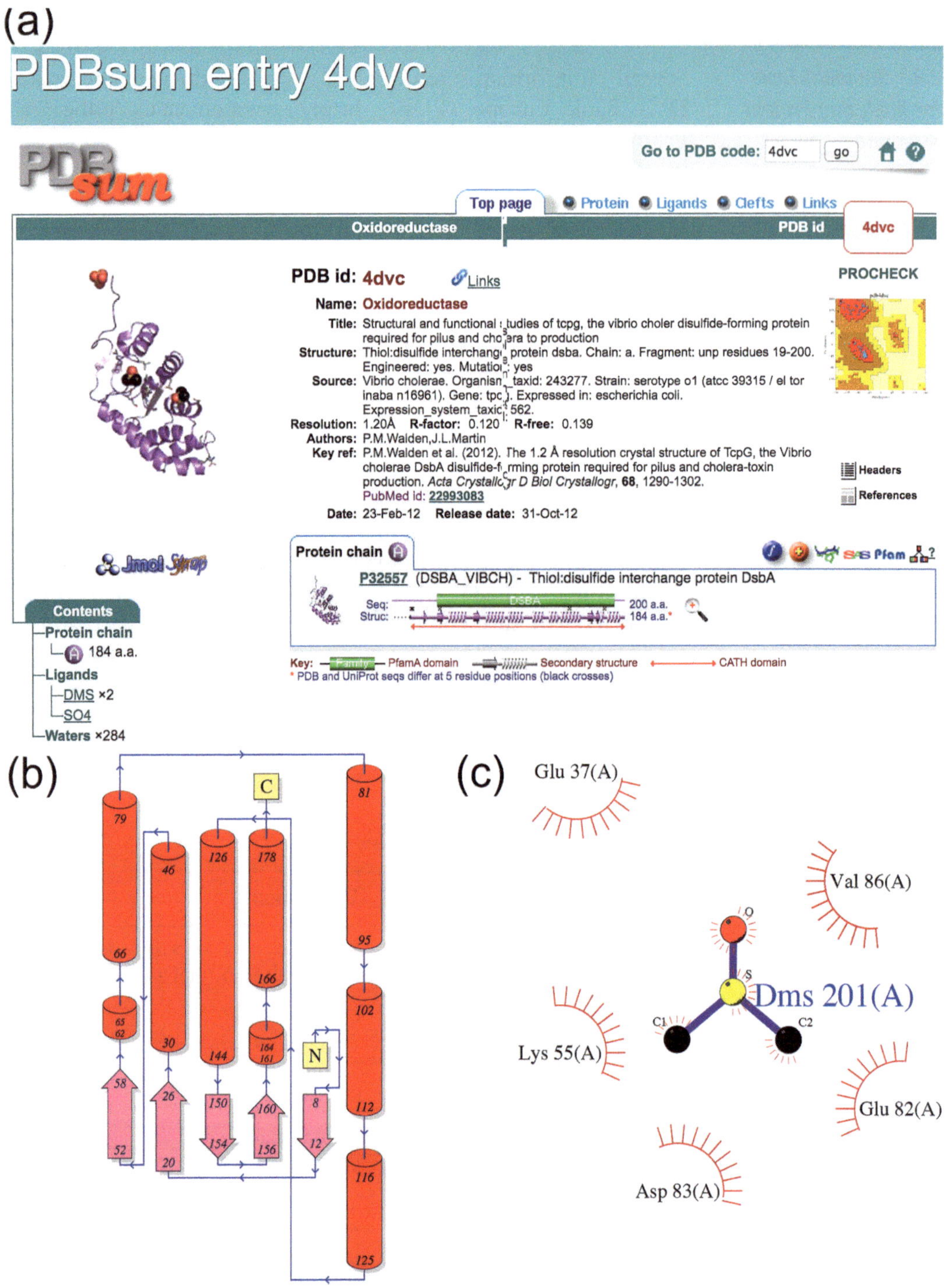

Fig. 2 Example from the PDBsum page for PDB entry 4DVC. (**a**) PDBsum summary showing the header information relating to the structure, molecular content, Pfam domain diagrams, links for viewing the coordinates in three-dimensions (Jmol) and to generate a comprehensive PROCHECK analysis. Additional links are also provided for databases like UniProtKB/Swiss-Prot, Pfam, SAS, and ArchSchema. The *protein*, *ligands*, and *cleft* tabs at the top of each PDBsum entry give access to topology diagrams, schematics depicting protein–protein/ligand/DNA interactions, and a description of the existing cleft, pores, and tunnels. (**b**) Topology diagram illustrating the secondary structure elements in 4DVC. *Numbers* correspond to the residues in the PDB and *arrows* indicate the direction of the protein chain. (**c**) LIGPLOT showing the interactions between the bound molecule (dimethyl sulfoxide) with the residues in the protein

Links are also provided to the protein sequence database UniProtKB/Swiss-Prot (http://www.uniprot.org/) and the Sequence annotated by Structure (SAS) tool (https://www.ebi.ac.uk/thornton-srv/databases/sas/), which scans the database and retrieves all related PDB structures and provides a multiple sequence alignment. Recently, a link to the ArchSchema program [21] (http://www.ebi.ac.uk/Tools/archschema) has been added to the PDBsum bioinformatics tool-set [16]. This program displays the Pfam domain architecture network that is most closely related to that of the query protein. The PDBsum output also provides topology diagrams showing the connectivity and relative positions of the secondary structure elements (Fig. 2b).

MAFFTash [22] (http://sysimm.ifrec.osaka-u.ac.jp/MAFFTash/) from the PDBj sever (http://pdbj.org) also performs multiple sequence alignments using sequence and structural information (Table 2).

2.2.2 Analysis of Protein Surface Properties

Bioinformatics tools are also available to analyze protein surface characteristics. For example PDBsum provides an analysis of all the grooves, pores, and tunnels in a given protein structure, which are computed using Mole 2.0 [23] and can be displayed through their respective tabs at the top of each PDBsum entry. In the same context, eF-surf (http://ef-site.hgc.jp/eF-surf/top.do) from PDBj is a web server that calculates the electrostatic potential and molecular surface of an uploaded file in pdb format. The server PDBePISA [24] is one of the most commonly used methods to analyze protein interfaces and predict quaternary structure (http://www.ebi.ac.uk/msd-srv/prot_int/pistart.html).

2.2.3 Analysis of Ligand-Bound Structures

When the PDB entry consists of DNA–protein or protein–ligand/metal complexes, PDBsum provides NUCPLOT [25] and LIGPLOT [26] diagrams, which are schematic representations of the DNA–protein and the ligand/metal–protein interactions, respectively (Fig. 2c). The PDBe also provides a number of complementary analysis resources and databases. DrugPort [16] (http://www.ebi.ac.uk/thornton-srv/databases/drugport/) (Table 2) is a tool that allows analyzing all the structural information in the PDB related to a query drug molecule. PDBeXpress (http://www.ebi.ac.uk/pdbe-srv/pdbexpress/) (Table 2) compiles a collection of tools that allows the identification of protein–ligand interactions from the PDB, along with searching for ligands that interact with a given set of residues or finding in the PDB all proteins that interact with a query ligand. Similarly, the PDBj eF-seek [27] (http://pdbj.org/help/ef-seek) can be used to survey the PDB to search for proteins with similar ligand binding sites as the probe PDB entry (Table 2).

2.3 Structural Bioinformatics Tools to Predict Protein Function

2.3.1 Predicting Function Based on Overall Structure

A large number of proteins deposited in biological databases do not share sequence similarity to any other known protein and consequently have uncharacterized functions. Typically, protein structures are more conserved than amino acid sequences and structural similarity can therefore be more informative with regard to defining protein function. A number of programs have been developed to identify structures in the PDB database that are globally similar to a given protein structure. Some of the most popular servers are DALI [28] (http://ekhidna.biocenter.helsinki.fi/dali_server), PDBeFold [29] (http://www.ebi.ac.uk/msd-srv/ssm/) and Structure Navigator [30] (http://pdbj.org/struc-navi) (Table 3).

Table 3
Selection of tools for structure comparison and function prediction

Server	URL	Description	Reference
DALI	http://ekhidna.biocenter.helsinki.fi/dali_server	Compares the 3D structure of a query protein against the whole PDB archive	[28]
PDBeFold	http://www.ebi.ac.uk/msd-srv/ssm/	Compares the 3D structure of a query protein against the whole PDB archive	[29]
Structure Navigator	http://pdbj.org/struc-navi	Compares the 3D structure of a query protein against the whole PDB archive	[30]
Cathedral	http://v3-4.cathdb.info/cgi-bin/CathedralServer.pl	For a given PDB probe identifies similar domains in the CATH database	[31]
PDBSiteScan	http://wwwmgs.bionet.nsc.ru/mgs/gnw/pdbsitescan/	Compares a query protein against known functional sites in the PDB	[33]
ProBiS	http://probis.cmm.ki.si/	Identifies proteins in the PDB with similar binding sites	[34]
PDBeMotif	http://www.ebi.ac.uk/pdbe-site/pdbemotif/	Finds conserved structural motifs and defines binding site characteristics	[35]
SiteEngine	http:/bioinfo3d.cs.tau.ac.il/SiteEngine	Identifies and compares binding sites and protein–protein interfaces	[36]
eF-Site	http://ef-site.hgc.jp/eF-site/	Calculates the electrostatic surface properties and identifies potential functional sites	[37]
ProFunc	https://www.ebi.ac.uk/thornton-srv/databases/profunc/	Identifies the likely biochemical function of a protein from its sequence, structure and/or binding site	[39]

2.3.2 Predicting Function Based on Structural Domains

If overall similarity to structurally characterized proteins cannot be obtained, the next approach to predict protein function is to identify similar domains, active sites and motifs which are evolutionary more conserved. The Cathedral server [31] (http://v3-4.cathdb.info/cgi-bin/CathedralServer.pl) allows the identification of similar domains from the CATH database [32]. PDBSiteScan [33] (http://wwwmgs.bionet.nsc.ru/mgs/gnw/pdbsitescan/) is used to identify active sites from PDBSite within the uploaded coordinate file of interest. Similarly, the more recent ProBiS database [34] (http://probis.cmm.ki.si/) allows the identification of proteins in the PDB repository that share similar binding sites. PDBeMotif [35] (http://www.ebi.ac.uk/pdbe-site/pdbemotif/) is an efficient tool that integrates protein sequence and protein structure in order to explore the PDB to find conserved structural motifs and protein–ligand interactions (Table 3).

2.3.3 Predicting Function Based on Physicochemical Properties

Other methods for predicting protein function also take into account the physicochemical properties of active sites and ligand binding sites. The server SiteEngine [36] (http:/bioinfo3d.cs.tau.ac.il/SiteEngine) uses the physicochemical properties of a ligand binding site to find similar sites in a query protein structure. The eF-Site database [37] (http://ef-site.hgc.jp/eF-site/) calculates the electrostatic and hydrophobic surface properties of query protein structures, and identifies potential active or ligand binding sites via a comparison to a database of known protein functional sites.

2.3.4 Predicting Function Based on Ligand Binding

An extension of ProBiS is the web server ProBiS-ligands [38] (http://probis.cmm.ki.si/ligands), which predicts the binding of ligands by superimposing similar ligand-bound PDB entries with the protein under investigation. Upon finding suitable fits the ligands are transposed onto the query protein. This tool could also be of interest in drug repurposing where new target proteins are identified for well-established drug molecules.

2.3.5 Predicting Function Based on Multiple Methods

Finally, the ProFunc server [39] (https://www.ebi.ac.uk/thornton-srv/databases/profunc/) combines a number of programs into the following streams that includes (1) primary amino acid sequence analysis, (2) fold and structural motif analysis along with (3) analysis against known ligand binding and active sites in order to identify possible sets of functionally related proteins to the query pdb. The output summary provides an at-a-glance view of the results from the different analyses performed.

3 Notes

Amino acid sequences and three-dimensional structures alone provide limited information about a protein. However, computational databases and tools allow researchers to validate and compare this information with hundreds of proteins of known structure and function, to obtain further information about the data quality, classification, function and evolution of their uncharacterised proteins. This process is cyclic whereby the sequence and structural data along with the information obtained from bioinformatics tools are fed back into the biological databases expanding their content. Despite the assistance provided by bioinformatics tools on the WWW, the success of this system is largely dependent on the users who are ultimately responsible for the accuracy of the information deposited in these public resources. Lastly it is critical to cite the bioinformatics tools that have played an integral part of this research so that their contribution is acknowledged and the development of these databases and programs can continue.

Acknowledgments

This work was supported by Australian Research Council (ARC) grant (DP150102287) and the Australian National Health and Medical Research Council (NHMRC) grant (APP1099151). B.H. is supported by an Australian Research Council Future Fellowship (FT130100580).

References

1. Reddy TB, Thomas AD, Stamatis D, Bertsch J, Isbandi M, Jansson J, Mallajosyula J, Pagani I, Lobos EA, Kyrpides NC (2015) The Genomes OnLine Database (GOLD) v. 5: a metadata management system based on a four level (meta)genome project classification. Nucleic Acids Res 43:D1099–D1106
2. UniProt C (2015) UniProt: a hub for protein information. Nucleic Acids Res 43:D204–D212
3. Bernstein FC, Koetzle TF, Williams GJ, Meyer EF Jr, Brice MD, Rodgers JR, Kennard O, Shimanouchi T, Tasumi M (1977) The Protein Data Bank: a computer-based archival file for macromolecular structures. J Mol Biol 112:535–542
4. Laskowski RA, Thornton JM (2008) Understanding the molecular machinery of genetics through 3D structures. Nat Rev Genet 9:141–151
5. Brunger AT (1992) Free R value: a novel statistical quantity for assessing the accuracy of crystal structures. Nature 355:472–475
6. Kleywegt GJ, Harris MR, Zou JY, Taylor TC, Wahlby A, Jones TA (2004) The Uppsala electron-density server. Acta Crystallogr D Biol Crystallogr 60:2240–2249
7. Laskowski RA, MacArthur MW, Moss DS, Thornton JM (1993) PROCHECK: a program to check the stereochemical quality of protein structures. J Appl Crystallogr 26:283–291
8. Vriend G (1990) WHAT IF: a molecular modeling and drug design program. J Mol Graph 8(52–56):29
9. Chen VB, Arendall WB 3rd, Headd JJ, Keedy DA, Immormino RM, Kapral GJ, Murray LW, Richardson JS, Richardson DC (2010) MolProbity: all-atom structure validation for macromolecular crystallography. Acta Crystallogr D Biol Crystallogr 66:12–21
10. Richardson JS, Schneider B, Murray LW, Kapral GJ, Immormino RM, Headd JJ, Richardson DC, Ham D, Hershkovits E, Williams LD, Keating KS, Pyle AM, Micallef D, Westbrook J, Berman HM, Consortium RNNO (2008) RNA backbone: consensus all-angle conformers and modular string nomenclature (an RNA Ontology Consortium contribution). RNA 14:465–481
11. Gore S, Velankar S, Kleywegt GJ (2012) Implementing an X-ray validation pipeline for the Protein Data Bank. Acta Crystallogr D Biol Crystallogr 68:478–483

12. Andrejasic M, Praaenikar J, Turk D (2008) PURY: a database of geometric restraints of hetero compounds for refinement in complexes with macromolecular structures. Acta Crystallogr D Biol Crystallogr 64:1093–1109
13. Weichenberger CX, Pozharski E, Rupp B (2013) Visualizing ligand molecules in Twilight electron density. Acta Crystallogr Sect F Struct Biol Cryst Commun 69:195–200
14. Lutteke T, von der Lieth CW (2004) pdb-care (PDB carbohydrate residue check): a program to support annotation of complex carbohydrate structures in PDB files. BMC Bioinformatics 5:69
15. Joosten RP, Salzemann J, Bloch V, Stockinger H, Berglund AC, Blanchet C, Bongcam-Rudloff E, Combet C, Da Costa AL, Deleage G, Diarena M, Fabbretti R, Fettahi G, Flegel V, Gisel A, Kasam V, Kervinen T, Korpelainen E, Mattila K, Pagni M, Reichstadt M, Breton V, Tickle IJ, Vriend G (2009) PDB_REDO: automated re-refinement of X-ray structure models in the PDB. J Appl Crystallogr 42:376–384
16. de Beer TA, Berka K, Thornton JM, Laskowski RA (2014) PDBsum additions. Nucleic Acids Res 42:D292–D296
17. Sayle RA, Milner-White EJ (1995) RASMOL: biomolecular graphics for all. Trends Biochem Sci 20:374
18. Finn RD, Mistry J, Schuster-Bockler B, Griffiths-Jones S, Hollich V, Lassmann T, Moxon S, Marshall M, Khanna A, Durbin R, Eddy SR, Sonnhammer EL, Bateman A (2006) Pfam: clans, web tools and services. Nucleic Acids Res 34:D247–D251
19. Hutchinson EG, Thornton JM (1996) PROMOTIF—a program to identify and analyze structural motifs in proteins. Protein Sci 5:212–220
20. Sillitoe I, Lewis TE, Cuff A, Das S, Ashford P, Dawson NL, Furnham N, Laskowski RA, Lee D, Lees JG, Lehtinen S, Studer RA, Thornton J, Orengo CA (2015) CATH: comprehensive structural and functional annotations for genome sequences. Nucleic Acids Res 43:D376–D381
21. Tamuri AU, Laskowski RA (2010) ArchSchema: a tool for interactive graphing of related Pfam domain architectures. Bioinformatics 26:1260–1261
22. Katoh K, Misawa K, Kuma K, Miyata T (2002) MAFFT: a novel method for rapid multiple sequence alignment based on fast Fourier transform. Nucleic Acids Res 30:3059–3066
23. Berka K, Hanak O, Sehnal D, Banas P, Navratilova V, Jaiswal D, Ionescu CM, Svobodova Varekova R, Koca J, Otyepka M (2012) MOLEonline 2.0: interactive web-based analysis of biomacromolecular channels. Nucleic Acids Res 40:W222–W227
24. Krissinel E, Henrick K (2007) Inference of macromolecular assemblies from crystalline state. J Mol Biol 372:774–797
25. Luscombe NM, Laskowski RA, Thornton JM (1997) NUCPLOT: a program to generate schematic diagrams of protein-nucleic acid interactions. Nucleic Acids Res 25: 4940–4945
26. Wallace AC, Laskowski RA, Thornton JM (1995) LIGPLOT: a program to generate schematic diagrams of protein-ligand interactions. Protein Eng 8:127–134
27. Kinoshita K, Murakami Y, Nakamura H (2007) eF-seek: prediction of the functional sites of proteins by searching for similar electrostatic potential and molecular surface shape. Nucleic Acids Res 35:W398–W402
28. Holm L, Rosenstrom P (2010) Dali server: conservation mapping in 3D. Nucleic Acids Res 38:W545–W549
29. Krissinel E, Henrick K (2004) Secondary-structure matching (SSM), a new tool for fast protein structure alignment in three dimensions. Acta Crystallogr D Biol Crystallogr 60:2256–2268
30. Standley DM, Kinjo AR, Kinoshita K, Nakamura H (2008) Protein structure databases with new web services for structural biology and biomedical research. Brief Bioinform 9:276–285
31. Redfern OC, Harrison A, Dallman T, Pearl FM, Orengo CA (2007) CATHEDRAL: a fast and effective algorithm to predict folds and domain boundaries from multidomain protein structures. PLoS Comput Biol 3, e232
32. Cuff AL, Sillitoe I, Lewis T, Clegg AB, Rentzsch R, Furnham N, Pellegrini-Calace M, Jones D, Thornton J, Orengo CA (2011) Extending CATH: increasing coverage of the protein structure universe and linking structure with function. Nucleic Acids Res 39:D420–D426
33. Ivanisenko VA, Pintus SS, Grigorovich DA, Kolchanov NA (2004) PDBSiteScan: a program for searching for active, binding and posttranslational modification sites in the 3D structures of proteins. Nucleic Acids Res 32:W549–W554
34. Konc J, Cesnik T, Konc JT, Penca M, Janezic D (2012) ProBiS database: precalculated binding site similarities and local pairwise alignments of PDB structures. J Chem Inf Model 52:604–612
35. Leontovich AM, Tokmachev KY, van Houwelingen HC (2008) The comparative analysis of statistics, based on the likelihood

ratio criterion, in the automated annotation problem. BMC Bioinformatics 9:31

36. Shulman-Peleg A, Nussinov R, Wolfson HJ (2005) SiteEngines: recognition and comparison of binding sites and protein-protein interfaces. Nucleic Acids Res 33:W337–W341
37. Kinoshita K, Nakamura H (2005) Identification of the ligand binding sites on the molecular surface of proteins. Protein Sci 14:711–718
38. Konc J, Janezic D (2014) ProBiS-ligands: a web server for prediction of ligands by examination of protein binding sites. Nucleic Acids Res 42:W215–W220
39. Laskowski RA, Watson JD, Thornton JM (2005) ProFunc: a server for predicting protein function from 3D structure. Nucleic Acids Res 33:W89–W93
40. Walden PM, Heras B, Chen KE, Halili MA, Rimmer K, Sharma P, Scanlon MJ, Martin JL (2012) The 1.2 A resolution crystal structure of TcpG, the Vibrio cholerae DsbA disulfide-forming protein required for pilus and cholera-toxin production. Acta Crystallogr D Biol Crystallogr 68:1290–1302

Chapter 17

In Silico Approach to Identify Potential Inhibitors for Axl-Gas6 Signaling

Swathik Clarancia Peter, Jayakanthan Mannu, and Premendu P. Mathur

Abstract

Axl-Gas6 signaling plays an important role in numerous cancers. Axl kinase, a member of receptor tyrosine kinase family is activated by different mechanisms with Gas6 as its major activator. Targeting the Axl with inhibitors may block the binding of Gas6 and further hinders the activation of Axl. This in turn inhibits the Axl-Gas6 signaling. Thus, inhibitors of the Axl kinase may serve as ideal drug candidates for treating many human cancers. In this study we carried out virtual screening of drug-like molecules from ZINC database to identify potential inhibitors for Axl kinase. Our virtual screening study showed that ZINC83758120, ZINC34079369, and ZINC83758121 are potential drug-like lead molecules to inhibit Axl kinase.

Key words Axl kinase docking, Gas6 protein, Zinc database,, Virtual screening, QikProp Glide docking

1 Introduction

Axl kinase is found to be overexpressed in many cancers like lung [1–3], breast [4, 5], prostrate [6], gastric [7], ovarian [8], and thyroid [9]. It is also found to be overexpressed in hepatocellular leukemia and acute myeloid leukemia [10, 11]. The level of Axl expression is comparatively high in cancer tissues to normal tissues [12]. The activated Axl kinase induces many signaling pathways involved in cell proliferation [12], metastasis [13], and inhibition of apoptosis [14, 15] by downstream signaling. Similarly, Gas6, a major ligand of Axl protein, has been reported for overexpression in many human cancers [16]. Both overexpression of Axl and over-activation of Axl-Gas6 signaling lead to poor prognosis [3, 17], and also correlated with therapeutic resistance [18, 19]. Hence, in this study, we have carried out virtual screening of lead-like molecules to identify potential compounds to inhibit Axl-Gas6 signaling pathway.

Shivakumar Keerthikumar and Suresh Mathivanan (eds.), *Proteome Bioinformatics*, Methods in Molecular Biology, vol. 1549, DOI 10.1007/978-1-4939-6740-7_17, © Springer Science+Business Media LLC 2017

Axl kinases are proteins belonging to the family of receptor tyrosine kinases (RTKs) which play important roles in many cancers and pathological conditions [20]. Axl signaling also has important roles in platelet function, spermatogenesis, and immunity. This protein consists of two immunoglobulin-like (IG) domains and two fibronectin type III domains (FNIII) in the extracellular region, a transmembrane domain, and a kinase domain in the cytoplasmic region [21, 22]. The activation of Axl kinase takes place by different mechanisms such as ligand-dependent dimerization, ligand-independent dimerization, hetero-dimerization with non-TAM receptor, and dimerization with the monomers on the neighboring molecules, of which, Gas6 (Growth Arrest Specific 6) is considered as the major and unique activator of Axl kinase by ligand-dependent dimerization mechanism. The protein structure of Gas6 contains a γ-carboxyglutamic acid [13] domain, loop region, four EGF-like repeats, and two C-terminal globular laminin G-like [19] domains. The binding activity of Axl-IG with Gas6-LG occurs at two sites, one being the major contact and the other being minor. It is retained by the Axl fragment consisting of two N-terminal immunoglobulin-like domains (Axl-IG) and LG1 domain of Gas6 [22]. Binding of Gas6 activates Axl and homodimerization of the molecule takes place which leads to tyrosine autophosphorylation and phosphorylation of downstream targets [21].

Identifying potential inhibitors which block Axl-Gas6 signaling axis may rectify the aberrant Axl signaling and can be ideal drug candidates to treat many types of cancers, thereby reducing the poor prognosis, decreasing the progression and invasiveness of the disease, and also increasing the drug sensitivity and efficacy.

2 Methods

2.1 Tools Used

1. Schrodinger Maestro 9.2.
2. Ligand library containing 7750 chemical compounds downloaded from ZINC database.

2.2 Protein Preparation

1. The experimental protein complex structure of Axl-Gas6 was retrieved from the Protein Data bank (http://www.rcsb.org/pdb/home/home.do) (PDB ID: 2C5D).
2. The retrieved protein complex structure was subjected to protein preparation using Maestro 9.3 protein preparation wizard in Schrodinger (*see* **Note 1**).
3. The protein complex was preprocessed by assigning bond orders and by adding hydrogen atoms.
4. Zero-order bonds were created for metal atoms. Disulphide bonds were created between Sulfur atoms that are within the range of 3.2 Å.

5. The water molecules beyond 5 Å from the hetero groups were deleted.
6. After preprocessing, the missing side chains were added using Prime module.

Residue	Type
A:389	ARG
A:413	GLU
B:389	ARG
B:413	GLU

7. During protein preparation, the hetero atoms of A:CA(1677), A:NAG-NAG, B:CA(1677), B:NAG-NAG, C:NI(1218), C:SO_4(1219), and D:NI(1218) were deleted.
8. The chains A and B of Gas6 were removed. The homologous chain D of Axl was removed.
9. Further, the protein structure was optimized for geometry to fix the orientations of thiols, hydroxyl, amides, histones.
10. The structure was optimized using PROPKA at the biological pH of 7.00.
11. The structure was minimized under the OPLS 2005 force field.

2.3 Grid Generation

After protein preparation, the grid at the site of active site was generated using Glide module in Schrodinger. It has been reported that mutation of Glu59 and Thr77 residues has dramatically reduces the binding of Axl with Gas6 [22]. Thus inhibitors binding to these residues can be ideal for inhibiting Axl-Gas6 binding, thereby preventing the activation of Axl receptor tyrosine kinase and its downstream signaling involved in oncogenic and pathological conditions. Here, we define abovementioned active site residues as centroid for grid generation.

1. The receptor-grid was generated with the centroid of the residues Glu59 and Thr77.
2. The van der Waal's radius scaling factor was set to 1.0 and the partial charge cutoff was set to 0.25. The charge scale factor was set to 1.0.

2.4 Ligand Preparation and Virtual Screening

1. The ligands in the input library were filtered based on ADMET properties using QikProp. The ligands were also pre-filtered by Lipinski's rule.
2. Ligands with reactive functional groups were removed. The input geometries of the ligands were regularized by epik.
3. The number of low energy conformations generated per ligand was one.

4. The virtual screening was carried out in Glide HTVS, Glide SP, and Glide XP under OPLS force field for ideal screening and docking of ligands at the binding site of the protein. Three poses were generated for each docked compound.

3 Results and Discussion

Axl kinase has been reported as a valid therapeutic target for many cancers [21]. The availability of three-dimensional structures of any target proteins plays a major role in designing inhibitors through computational approaches. In spite of experimental structure of Axl kinase has been reported in the year 2005 [22], so far no attempt to find for potential inhibitors through computational approaches. Here, we have attempted virtual screening of lead like chemical molecules from ZINC database using virtual screening workflow of Schrodinger suite 2012. The virtual screening of chemical library comprising 7750 compounds against the protein Axl kinase identified three ligands with optimal binding free energy (*see* **Notes 2–4**). These ligands are ZINC83758120 (2-[[(1R)-2-amino-1-(5-bromo-2-furyl)ethyl]amino]ethanol), ZINC34079369 ((1R)-2-(2-aminoethylamino)-1-(2,6-dichlorophenyl)ethanol), and ZINC83758121 (2-[[(1S)-2-amino-1-(5-bromo-2-furyl)ethyl]amino]ethanol).

3.1 ZINC83758120

The virtual screening of ZINC database produced ZINC83758120 as a top scoring ligand molecule with binding free energy of −44.074 kcal/mol. Analysis of interaction pattern of this compound shows that four hydrogen bonds were formed by amino residues of Axl kinase. In which, two bonds were formed with Gln78 residue (hydrogen bond distance of 3.01 and 2.84 Å), one with Glu56 (hydrogen bond distance of 2.82 Å) and another with Glu85 (hydrogen bond distance of 2.61 Å) (Table 1). In addition, the binding of ZINC83758120 with Axl kinase was also further stabilized by van der Waal's interactions by amino residues such as Trp89, Glu85, Gln78, and Glu56 at the scaling factor of 1.00 Å (Fig. 1A (a, b)).

3.2 ZINC34079369

The second top scored ligand molecule was ZINC34079369 with binding free energy of −35.167 kcal/mol. This compound formed two hydrogen bonds with Axl kinase amino acid residues such as Gln76 and Ser74. The side chain nitrogen atom of Gln76 acts as a hydrogen bond donor to form hydrogen bond with oxygen atom of this drug-like molecule at a distance of 3.11 Å. Another hydrogen bond was formed between oxygen atom of this drug-like molecule and side chain oxygen atom of Ser74 at a distance of 2.75 Å (Table 1). The residues which formed van der Waal's interactions are Ser74, Ala72, Glu70, Leu69, and Glu59 at the scaling factor of 1.00 Å (Fig. 1B (a, b)).

Table 1
Molecular interactions of lead-like molecules with Axl kinase

Lead-like molecules	Hydrogen bond donor	Hydrogen bond acceptor	Hydrogen bond length (Å)	VdW interaction residues (scaling factor = 1.00 Å)	Glide energy (glide emodel) (kcal/mol)
ZINC83758120	Lead2:N1 GLN78:NE2 Lead2:O2 Lead2:N1	GLU85:OE1 LEAD2:O2 GLN78:OE1 GLU56:OE1	2.61 3.01 2.84 2.82	TRP89, GLU85, GLN78, GLU56	−44.074
ZINC34079369	GLN76:NE2 Lead1:O1	Lead1:O1 SER74:OG	3.11 2.75	SER74, ALA72, GLU70, LEU69, GLU59	−35.167
ZINC83758121	GLN78:NE2 Lead3:O2 Lead3:N1	Lead3:O2 PRO57:O GLU56:OE1	2.67 2.97 2.55	TRP89, GLN78, PRO57, GLU56	−34.833

3.3 ZINC83758121

The third top scored screened ligand molecule wasZINC83758121. This compound showed a binding energy of −34.833 kcal/mol with Axl kinase. ZINC83758121 formed three hydrogen bonds with the residues Glu56, Pro57, and Gln78 of Axl at the distance of 2.55, 2.97, and 2.67 Å respectively (Table 1). The residues with van der Waal's interactions at the scaling factor of 1.00 Å are Glu56, Pro57, Gln78, and Trp89 (Fig. 1C (a, b)).

Our virtual screening results showed that ZINC83758120 forms a stable interaction with the Axl kinase protein. This compound also showed comparatively strong interactions within the cavity of Axl kinase from other two lead molecules such as ZINC34079369 and ZINC83758121 in terms of number of hydrogen bonds and binding free energy.

4 Conclusion

Our study showed that ZINC83758120 would be a potential lead like molecule to design inhibitors for Axl kinase. This compound can be further improved by modifying the existing groups or introducing new chemical moiety to enhance its binding affinity and thereby increasing the efficacy of the compound.

5 Notes

1. The initial requirement of any docking study is the availability of protein structure for the target protein. The experimental protein complex structure of Axl-Gas6 was retrieved from the

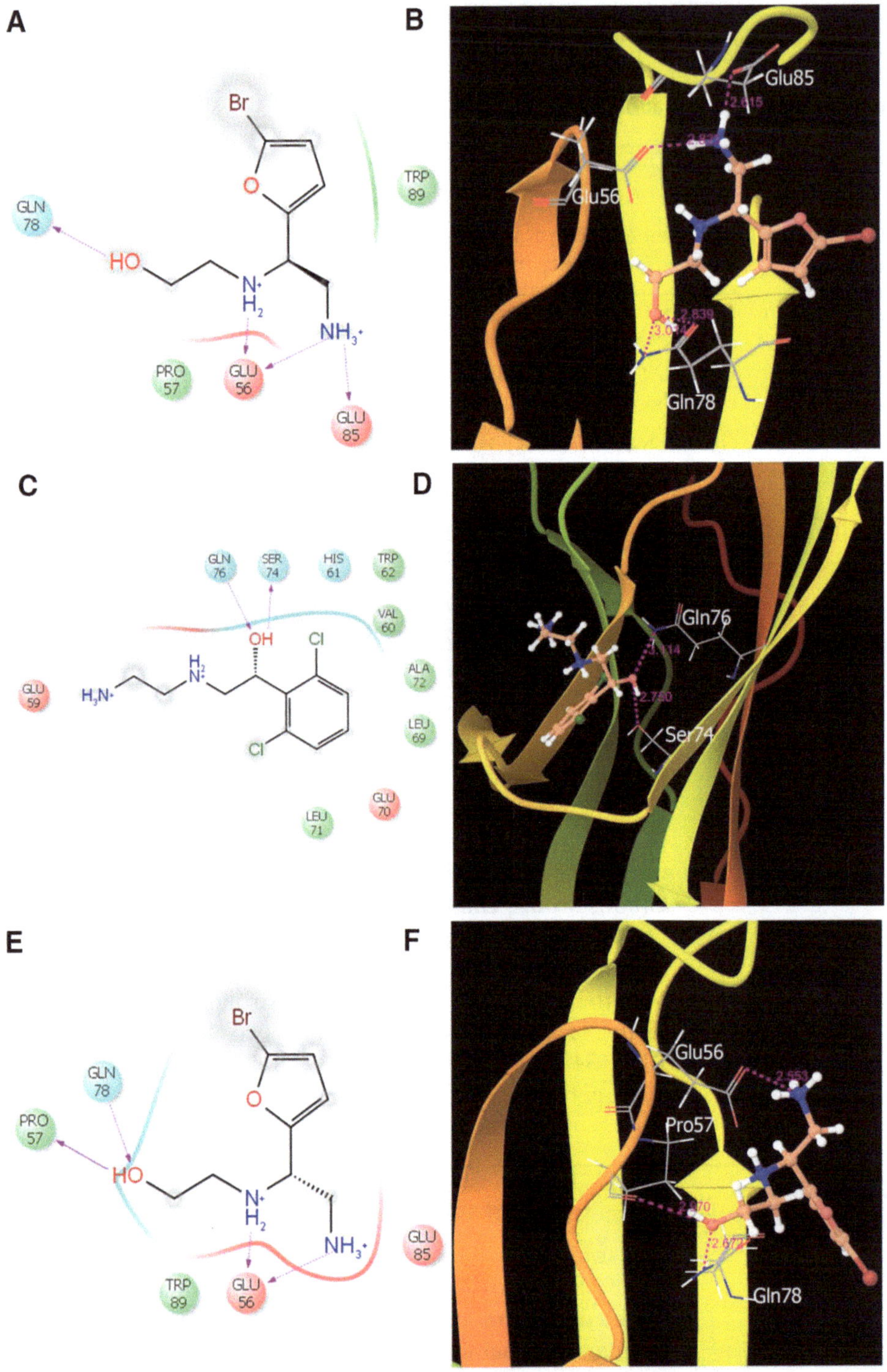

Fig. 1 The docked complexes of ZINC83758120 (**A**, **B**), ZINC34079369 (**C, D**), and ZINC83758121 (**E, F**). **A, C** and **E** represents two-dimensional view of the docked complexes. The interacting residues are represented in spheres (*Red*: negatively charged residues; *Violet*: positively charged residues; *Cyan*: polar residues; *Green*: hydrophobic residues; *Pink color dashed arrow*: hydrogen bonds. **B, D** and **F** represents three-dimensional view of the docked complex. The interacting amino residues are represented in line model and colored by atom types (*Grey*: carbon; *white*: hydrogen, *red*: oxygen; *blue*: nitrogen). The interacting ligand represented in ball and stick model

Protein Data bank (http://www.rcsb.org/pdb/home/home.do) (PDB ID: 2C5D). This co-crystallized structure consists of four chains namely A, B, C, and D, in which chains A and B belong to Gas6 and C and D chains belong to Axl kinase. The chains A, B of Gas6 were removed. Since the chain C and D were homologous, it is sufficient to perform screening for any one of the two chains, so the chain D of Axl was removed. The docking and virtual screening was carried out only for chain C of Axl kinase.

2. Chemical library consisting of lead-like compounds was obtained from ZINC database (http://zinc.docking.org/). ZINC database is a free database of commercially available compounds. We have retrieved a total of 7750 compounds from this database and it was used for further virtual screening. There are many other databases from which the library of chemical compounds can be downloaded.
3. The restrain minimization is performed to remove atom clashes and to relax side chains.
4. The grid can be generated either by selecting the residues in amino acid sequence of the protein or by specifying the X, Y, and Z coordinates of the residues around which the grid has to be generated.

Acknowledgment

Research in the laboratory of Bioinformatics, Tamil Nadu Agricultural University is supported by BTIS scheme of Department of Biotechnology (DBT), Government of India, New Delhi, India.

Key Terms and Definitions

Axl kinase	Axl kinase is an enzyme of receptor tyrosine kinase subfamily.
Docking	A method used to predict molecular interactions between two molecules. These molecules are protein, DNA, and small molecules.
GAS6 protein	Gas6, a major ligand of Axl protein, has been reported for overexpression in many human cancers.
Zinc database	A free database of chemical compounds for virtual screening. This database contains over 35 million compounds to be used for virtual screening.

Virtual screening	A method of predicting interactions of small molecules from a library of compounds against a cavity of target protein structures.
QIKPROP	A modules of Schrodinger Maestro 9.3 program, which could be used to identify drug toxicity of compounds.
Glide docking	A modules of Schrodinger Maestro 9.3 program, which could be used for molecular docking.

References

1. Shieh YS, Lai CY, Kao YR, Shiah SG, Chu YW, Lee HS, Wu CW (2005) Expression of axl in lung adenocarcinoma and correlation with tumor progression. Neoplasia 7(12): 1058–1064
2. Verma A, Warner SL, Vankayalapati H, Bearss DJ, Sharma S (2011) Targeting Axl and Mer kinases in cancer. Mol Cancer Ther 10(10):1763–1773, doi:1535–7163.MCT-11-0116 [pii]10.1158/1535-7163.MCT-11-0116
3. Ishikawa M, Sonobe M, Nakayama E, Kobayashi M, Kikuchi R, Kitamura J, Imamura N, Date H (2013) Higher expression of receptor tyrosine kinase Axl, and differential expression of its ligand, Gas6, predict poor survival in lung adenocarcinoma patients. Ann Surg Oncol 20(Suppl 3):S467–S476. doi:10.1245/s10434-012-2795-3
4. Berclaz G, Altermatt HJ, Rohrbach V, Kieffer I, Dreher E, Andres AC (2001) Estrogen dependent expression of the receptor tyrosine kinase axl in normal and malignant human breast. Ann Oncol 12(6):819–824
5. Meric F, Lee WP, Sahin A, Zhang H, Kung HJ, Hung MC (2002) Expression profile of tyrosine kinases in breast cancer. Clin Cancer Res 8(2):361–367
6. Jacob AN, Kalapurakal J, Davidson WR, Kandpal G, Dunson N, Prashar Y, Kandpal RP (1999) A receptor tyrosine kinase, UFO/Axl, and other genes isolated by a modified differential display PCR are overexpressed in metastatic prostatic carcinoma cell line DU145. Cancer Detect Prev 23(4):325–332, doi:cdp99034 [pii]
7. Wu CW, Li AF, Chi CW, Lai CH, Huang CL, Lo SS, Lui WY, Lin WC (2002) Clinical significance of AXL kinase family in gastric cancer. Anticancer Res 22(2B):1071–1078
8. Rankin EB, Fuh KC, Taylor TE, Krieg AJ, Musser M, Yuan J, Wei K, Kuo CJ, Longacre TA, Giaccia AJ (2010) AXL is an essential factor and therapeutic target for metastatic ovarian cancer. Cancer Res 70(19):7570–7579, doi:0008–5472.CAN-10-1267 [pii]10.1158/0008-5472.CAN-10-1267
9. Ito T, Ito M, Naito S, Ohtsuru A, Nagayama Y, Kanematsu T, Yamashita S, Sekine I (1999) Expression of the Axl receptor tyrosine kinase in human thyroid carcinoma. Thyroid 9(6):563–567
10. He L, Zhang J, Jiang L, Jin C, Zhao Y, Yang G, Jia L (2010) Differential expression of Axl in hepatocellular carcinoma and correlation with tumor lymphatic metastasis. Mol Carcinog 49(10):882–891. doi:10.1002/mc.20664
11. Hong CC, Lay JD, Huang JS, Cheng AL, Tang JL, Lin MT, Lai GM, Chuang SE (2008) Receptor tyrosine kinase AXL is induced by chemotherapy drugs and overexpression of AXL confers drug resistance in acute myeloid leukemia. Cancer Lett 268(2):314–324, doi:S0304-3835(08)00284-X [pii]10.1016/j.canlet.2008.04.017
12. Paccez JD, Vasques GJ, Correa RG, Vasconcellos JF, Duncan K, Gu X, Bhasin M, Libermann TA, Zerbini LF (2013) The receptor tyrosine kinase Axl is an essential regulator of prostate cancer proliferation and tumor growth and represents a new therapeutic target. Oncogene 32(6):689–698, doi:onc201289 [pii]10.1038/onc.2012.89
13. Gjerdrum C, Tiron C, Hoiby T, Stefansson I, Haugen H, Sandal T, Collett K, Li S, McCormack E, Gjertsen BT, Micklem DR, Akslen LA, Glackin C, Lorens JB (2009) Axl is an essential epithelial-to-mesenchymal transition-induced regulator of breast cancer metastasis and patient survival. Proc Natl Acad Sci U S A 107(3):1124–1129, doi:0909333107 [pii]10.1073/pnas.0909333107
14. van Ginkel PR, Gee RL, Shearer RL, Subramanian L, Walker TM, Albert DM, Meisner LF, Varnum BC, Polans AS (2004)

Expression of the receptor tyrosine kinase Axl promotes ocular melanoma cell survival. Cancer Res 64(1):128–134

15. Wilhelm I, Nagyoszi P, Farkas AE, Couraud PO, Romero IA, Weksler B, Fazakas C, Dung NT, Bottka S, Bauer H, Bauer HC, Krizbai IA (2008) Hyperosmotic stress induces Axl activation and cleavage in cerebral endothelial cells. J Neurochem 107(1):116–126, doi:JNC5590 [pii]10.1111/j.1471-4159.2008.05590.x
16. Mc Cormack O, Chung WY, Fitzpatrick P, Cooke F, Flynn B, Harrison M, Fox E, Gallagher E, Goldrick AM, Dervan PA, Mc Cann A, Kerin MJ (2008) Growth arrest-specific gene 6 expression in human breast cancer. Br J Cancer 98(6):1141–1146, doi:6604260 [pii]10.1038/sj.bjc.6604260
17. Hutterer M, Knyazev P, Abate A, Reschke M, Maier H, Stefanova N, Knyazeva T, Barbieri V, Reindl M, Muigg A, Kostron H, Stockhammer G, Ullrich A (2008) Axl and growth arrest-specific gene 6 are frequently overexpressed in human gliomas and predict poor prognosis in patients with glioblastoma multiforme. Clin Cancer Res 14(1):130–138, doi:14/1/130 [pii]10.1158/1078-0432.CCR-07-0862
18. Bansal N, Mishra PJ, Stein M, DiPaola RS, Bertino JR (2015) Axl receptor tyrosine kinase is up-regulated in metformin resistant prostate cancer cells. Oncotarget 6(17):15321–15331, doi:4148 [pii]
19. Brand TM, Iida M, Stein AP, Corrigan KL, Braverman CM, Luthar N, Toulany M, Gill PS, Salgia R, Kimple RJ, Wheeler DL (2014) AXL mediates resistance to cetuximab therapy. Cancer Res 74(18):5152–5164, doi:0008-5472.CAN-14-0294 [pii]10.1158/0008-5472.CAN-14-0294
20. O'Donnell K, Harkes IC, Dougherty L, Wicks IP (1999) Expression of receptor tyrosine kinase Axl and its ligand Gas6 in rheumatoid arthritis: evidence for a novel endothelial cell survival pathway. Am J Pathol 154(4):1171–1180, doi:S0002-9440(10)65369-2 [pii]10.1016/S0002-9440(10)65369-2
21. Axelrod H, Pienta KJ (2014) Axl as a mediator of cellular growth and survival. Oncotarget 5(19):8818–8852, doi:2422 [pii]
22. Sasaki T, Knyazev PG, Clout NJ, Cheburkin Y, Gohring W, Ullrich A, Timpl R, Hohenester E (2006) Structural basis for Gas6-Axl signalling. EMBO J 25(1):80–87, doi:7600912 [pii]10.1038/sj.emboj.7600912

INDEX

Shivakumar Keerthikumar and Suresh Mathivanan (eds.), *Proteome Bioinformatics*, Methods in Molecular Biology, vol. 1549, DOI 10.1007/978-1-4939-6740-7, © Springer Science+Business Media LLC 2017

MIX
Papier aus verantwortungsvollen Quellen
Paper from responsible sources
FSC® C105338

If you have any concerns about our products,
you can contact us on
ProductSafety@springernature.com

In case Publisher is established outside the EU,
the EU authorized representative is:
Springer Nature Customer Service Center GmbH
Europaplatz 3, 69115 Heidelberg, Germany

Printed by Libri Plureos GmbH
in Hamburg, Germany